Romeo Bianchetti

Control and readout of a superconducting artificial atom

Romeo Bianchetti

Control and readout of a superconducting artificial atom

Cavity quantum electrodynamics in high quality transmission line resonators with superconducting qubits

Südwestdeutscher Verlag für Hochschulschriften

Impressum/Imprint (nur für Deutschland/only for Germany)
Bibliografische Information der Deutschen Nationalbibliothek: Die Deutsche Nationalbibliothek verzeichnet diese Publikation in der Deutschen Nationalbibliografie; detaillierte bibliografische Daten sind im Internet über http://dnb.d-nb.de abrufbar.
Alle in diesem Buch genannten Marken und Produktnamen unterliegen warenzeichen-, marken- oder patentrechtlichem Schutz bzw. sind Warenzeichen oder eingetragene Warenzeichen der jeweiligen Inhaber. Die Wiedergabe von Marken, Produktnamen, Gebrauchsnamen, Handelsnamen, Warenbezeichnungen u.s.w. in diesem Werk berechtigt auch ohne besondere Kennzeichnung nicht zu der Annahme, dass solche Namen im Sinne der Warenzeichen- und Markenschutzgesetzgebung als frei zu betrachten wären und daher von jedermann benutzt werden dürften.

Verlag: Südwestdeutscher Verlag für Hochschulschriften GmbH & Co. KG
Dudweiler Landstr. 99, 66123 Saarbrücken, Deutschland
Telefon +49 681 37 20 271-1, Telefax +49 681 37 20 271-0
Email: info@svh-verlag.de

Approved by: Zürich, ETH, Diss. No. 19174, 2010

Herstellung in Deutschland:
Schaltungsdienst Lange o.H.G., Berlin
Books on Demand GmbH, Norderstedt
Reha GmbH, Saarbrücken
Amazon Distribution GmbH, Leipzig
ISBN: 978-3-8381-1179-7

Imprint (only for USA, GB)
Bibliographic information published by the Deutsche Nationalbibliothek: The Deutsche Nationalbibliothek lists this publication in the Deutsche Nationalbibliografie; detailed bibliographic data are available in the Internet at http://dnb.d-nb.de.
Any brand names and product names mentioned in this book are subject to trademark, brand or patent protection and are trademarks or registered trademarks of their respective holders. The use of brand names, product names, common names, trade names, product descriptions etc. even without a particular marking in this works is in no way to be construed to mean that such names may be regarded as unrestricted in respect of trademark and brand protection legislation and could thus be used by anyone.

Publisher: Südwestdeutscher Verlag für Hochschulschriften GmbH & Co. KG
Dudweiler Landstr. 99, 66123 Saarbrücken, Germany
Phone +49 681 37 20 271-1, Fax +49 681 37 20 271-0
Email: info@svh-verlag.de

Printed in the U.S.A.
Printed in the U.K. by (see last page)
ISBN: 978-3-8381-1179-7

Copyright © 2011 by the author and Südwestdeutscher Verlag für Hochschulschriften GmbH & Co. KG and licensors
All rights reserved. Saarbrücken 2011

In Ts'ui Pên's novel,
all the outcomes in fact occur;
each is the starting point for further bifurcations.
Once in a while, the paths of that labyrinth converge:
for example, you come to this house,
but in one of the possible pasts you are my enemy, in another my friend.

- *Jorge Luis Borges*

Abstract

Quantum mechanics is an overwhelmingly successful theory, nevertheless nearly all of its applications do not demonstrate explicitly its most striking features. Even seminal devices such as lasers and transistors only imply quantum mechanics in a statistical sense, while all macroscopic observables can be described by classical theories. The realization of a true quantum machine could address problems which are intractable with any classical device such as the simulation of complex quantum systems or the efficient factorization of big numbers to cite only two examples.

Cavity quantum electrodynamics (CQED) features all unique characteristics of quantum mechanics, studying the strong interaction of single photons and atoms. The implementation of artificial superconducting atoms in high quality transmission line cavities realizes such a CQED setup in a solid state environment. This opens the field of quantum optics to embedded devices. Furthermore, the relatively simple fabrication of these devices makes them a promising candidate for the realization of a quantum information processor. In this architecture, macroscopic quantum circuits act as effective two- and three-level systems (qubits and qutrits) by employing the large non-linearity of a Josephson junction. They strongly couple to single photons in a one-dimensional superconducting cavity which inhibits radiative decay for the contained fragile states and at the same time acts as a readout device.

During this thesis a new laboratory for circuit cavity quantum electrodynamics experiments at cryogenic temperatures and high frequencies is set up at ETH Zurich. Devices fabricated at ETH are measured, demonstrating the classical signatures of CQED operated resonantly such as vacuum Rabi mode splitting, and dispersive effects such as AC-Stark and Lamb shifts.

The precise characterization of the relevant parameters of two- and three-level artificial atoms, combined with an accurate model of coherent and dissipative dynamics of the externally driven system show excellent quantitative agreement between data and theory. On this basis high fidelity measurements of the qutrit populations by monitoring the field transmitted through the cavity are demonstrated. Arbitrary coherent superposition states to up to three levels are prepared with high quality using optimal control techniques. They are characterized for the first time outside the field of photon optics by a tomographic method. Full three-level quantum state tomography enables to test simplified qubit algorithms or generalized Bell-inequalities and can be extended to several coupled systems.

Finally the qubit-induced nonlinear cavity response is analyzed in the dispersive regime and used as a measurement device. In the high-power, non-linear regime high fidelity single-shot qubit read-out can readily be implemented without the need of additional devices.

Zusammenfassung

Trotz der überwältigende erfolge der Quantenmechanik, zeigt fast keine Anwendung ihre bemerkenswertesten Eigenschaften. Selbst wegweisende Technologien wie Laser oder Transistoren setzen die Quantenmechanik nur im statistischen Sinne ein, während alle makroskopischen Observablen mit klassischen Theorien beschrieben werden können. Die Realisierung einer echten Quantenmaschine könnte sich mit Problemen befassen die für klassische Computer schwer zu bewältigen sind, wie zum Beispiel die Simulation von komplexen Quantensystemen oder die effiziente Primfaktorzerlegung grosser Zahlen.

Hohlraumquantenelektrodynamik (CQED) untersucht die starke Wechselwirkung zwischen einzelnen Photonen und Atomen und weist alle frappierende Eigenschaften der Quantenmechanik auf. Mittels künstlicher Atome in Mikrowellenresonatoren hoher Güte, kann die Hohlraumquantenelektrodynamik in einem Festkörper realisiert werden. Diese Konstruktion eröffnet einem das Feld der Quantenoptik in integrierten Schaltungen. Ausserdem stellt sie wegen der relativ einfachen Herstellung solcher Bauelemente einen vielversprechenden Anwärter für die Implementierung eines Quanteninformationsprozessors dar. In dieser Architektur fungieren makroskopische Quantenschaltungen als effektive Zwei- und Dreiniveausysteme (Qubits und Qutrits) indem sie die grosse Nichtlinearität von Josephson-Kontakten ausnutzen und stark zu einzelnen Photonen in einem supraleitenden quasi eindimensionalen Hohlraumresonatoren koppeln. Dieser unterbindet strahlungsbedingte Zerfälle von den fragilen Zuständen und dient gleichzeitig als Messinstrument.

Im Rahmen der vorliegenden Dissertationsarbeit wurde an der ETH Zürich ein neues Tieftemperatur und Hochfrequenzlabor aufgebaut um Experimente im Feld der Hohlraumquantenelektrodynamik durchzuführen. An der ETH hergestellte Proben wurden charakterisiert und klassische Signaturen der resonanten CQED, wie die Vakuum-Rabi-Modenaufspaltung, und dispersive Effekte, wie die Lamb- und Starkverschiebung, nachgewiesen.

Die gemessenen Daten sind in exzellenter quantitativer Übereinstimmung mit einem Modell für das getriebene System, das kohärente und dissipative Dynamik einbezieht. Die Population des Qutrits wird dabei durch der Beobachtung des Feldes, das durch den Resonator transmittiert wird rekonstruiert. Beliebige kohärente Überlagerungen von bis zu drei Zuständen konnten dank der Methoden der optimalen Steuerung mit hoher Güte präpariert werden. Diese wurden erstmals ausserhalb des Feldes der Photonenoptik mit einer tomographischen Methode charakterisiert. Volle Dreiniveauquantenzustandstomographie ermöglicht das Testen vereinfachter Qubitalgorithmen oder verallgemeinerter Bell-Ungleichungen und kann auf mehrere gekoppelte Systeme erweitert werden.

Schliesslich ist die Qubit-induzierte nichtlineare Resonatorantwort im dispersivem Regime analysiert und als Messvorrichtung benutzt worden. Im Bereich weit oberhalb der kritischen Leistung kann der Qubitzustand mit hoher Genauigkeit durch eine einzelne Messung ausgelesen werden.

Contents

List of Symbols   XI

1   Introduction   1
    1.1   Quantum information processing 2
    1.2   Quantum optics on a chip 4

2   Circuit Quantum Electrodynamics   7
    2.1   Transmission line cavities 8
    2.2   The Cooper-pair box 12
    2.3   Reaching the strong coupling regime of circuit QED 15
    2.4   Dispersive regime 18
    2.5   Transmon artificial atom 19
    2.6   Dispersive transmon regime 22

3   Experimental Setup   25
    3.1   Cryogenic wiring 25
    3.2   Noise temperature of the amplification chain 29
    3.3   Amplitude and phase modulation of the control signal 33
    3.4   Measurement signal demodulation 37
    3.5   Selective DC flux control 39
    3.6   Sample fabrication 44
    3.7   Power dependence of high Q resonators 45

4 Dispersive Quantum State Readout   **51**
    4.1 Cavity-Bloch equations 54
    4.2 Continuous measurement 56
    4.3 Pulsed measurement 61
    4.4 Population reconstruction 63
        4.4.1 Maximizing signal-to-noise ratio 65
    4.5 Rabi oscillations measurements 66
    4.6 Energy decoherence measurements 67
    4.7 T_2 measurements using Ramsey oscillations 69

5 Generalization to 3-levels   **71**
    5.1 Three level Cavity-Bloch equations 72
    5.2 3-level population reconstruction 76
    5.3 Optimization of the measurement frequency 77
    5.4 Rabi oscillations on the 2nd excited state 78
    5.5 Measurement of the Rabi rates 80
    5.6 Pulse optimization 81
    5.7 T_1-time of the 2nd excited state 86
    5.8 Phase coherence of the 2nd excited state 88
    5.9 3-level tomography 89
    5.10 Outlook 92

6 Nonlinear Cavity Response   **93**
    6.1 High power cavity response 93
    6.2 Nonlinearity versus detuning 96
    6.3 High power qubit read-out 99
    6.4 Single shot read-out 102

Appendices   **105**

A Microwave Devices at Cryogenic Temperatures   **105**
    A.1 Cryogenic heat flows 105
    A.2 One-dimensional black body radiation 107
    A.3 Circulators 107

	A.4 Copper powder filters	109
	A.5 Low noise power supply	110
	A.6 Switch electronics	111
B	**Numerical Recipes**	**113**
	B.1 Three level cavity-Bloch equations	113
	B.2 Maximum likelihood estimation	116

Bibliography   119

Acknowledgements   137

List of Publications   139

List of Symbols

$a^{(\dagger)}$	photon annihilation (creation) operator
c	speed of light in vacuum [m/s]
C	capacitance [F]
E_c	charging energy [J]
E_J	Josephson energy [J]
g_i	coupling constant of the transition $i, i+1$ to the field [Hz]
G	gain
k_B	Boltzmann constant [J/K]
h	Planck constant
I	in-phase quadrature
IL	insertion loss
$\hat{M}$	measurement operator
n	number of photons
n_{crit}	critical photon number
P	power [W]
p_i	population of state $|i\rangle$

Q_i	quality factor labeled by i
Q	out-of-phase quadrature
R	resistance [Ohm]
s_i	signal on channel i [V]
S_{ij}	scattering parameters of linear electrical networks
S_i	AC-stark shift from state $|i\rangle$ [MHz]
SNR	signal to noise ratio
V_{rms}^0	rms voltage between the center conductor and the ground planes [V]
T	temperature [K]
T_1^i	energy decay time of state $|i\rangle$ [s]
T_2^i	dephasing time of state $|i\rangle$ [s]
T_ϕ^i	pure dephasing time of state $|i\rangle$ [s]
T_n	noise temperature [K]
U	unitary matrix
Z	impedance [Ohm]
γ_1^i	energy decay rate of state $|i\rangle$ [Hz]
γ_2^i	dephasing rate of state $|i\rangle$ [Hz]
γ_ϕ^i	pure dephasing rate of state $|i\rangle$ [Hz]
$\Delta_{ij} = \omega_i - \omega_j$	detuning between tone i and j [Hz]
$\Delta_s^i = \Delta_{i,i+1} - \omega_s$	detuning between excitation tone ω_s and transition ij [Hz]
ϵ	electric permittivity [F/m]
ϵ_{eff}	effective permittivity
ϵ_m	relative permittivity
ϵ_m	drive amplitude, at frequency ω_m, in units of [Hz]

ϵ_{ij}	drive amplitude, at frequency ω_{ij}^s, in units of [Hz]
κ	photon decay rate [Hz]
λ	wavelength [m]
μ	magnetic permeability
μ_0	magnetic permeability of free space
μ_r	relative magnetic permeability
ρ	density matrix
σ_i	Pauli matrices
ϕ	magnetic flux
φ	phase
χ	dispersive resonator shift in the 2-level approximation [Hz]
χ_i	dispersive resonator shift from state $|i\rangle$ [Hz]
ω_i	transition frequency of state $|i\rangle$ [Hz]
$|i\rangle$	quantum state of the i-th excitation
$|i\rangle\langle j| = \hat{P}_{ij}$	projector of state j on state i
$\langle . \rangle$	expectation value
$\Im$	imaginary part
$\Re$	real part
Tr	trace
$\hat{.}$	operator

XIII

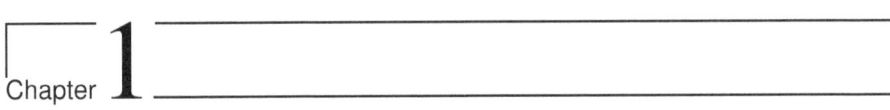

Chapter 1

Introduction

Information technology has become ubiquitous in everyday life as the need for processing and storing information of our society continues to grow. Semiclassical ensemble models can successfully describe the operation of semiconducting devices which are at the basis of modern computing where the striking features of quantum mechanics do not matter. Counterintuitive phenomena like tunneling, superposition of states or the uncertainty principle do never play a role. The outstanding success of integrated circuits is, however, based on the exponential miniaturization of the electronic structures, currently as small as 32 nm. The domain where few or single electrons carry information or are used to perform computations, thus where quantum mechanics must be fully taken into account is therefore not far away. This is not a small correction to the classical newtonian dynamics, but opens the doors to a new discipline which has outreach in regimes which are intractable with current tools. A priori security in communication [Gisin02], efficient simulation of complex quantum systems [Buluta09], enhancements in metrology [Uzan03, Giovannetti06] and exponential speedup of selected computational tasks [Spiller05] become possible.

The technological effort towards the ultimate control of isolated quantum systems is, however, not the only driving force in current research. Unanswered, fundamental questions about the nature of reality [Einstein35]
[Schrödinger35] and information processing capabilities [Turing37], dating back to the fathers of quantum mechanics and information theory can finally be experimentally tested. Furthermore, the observation of the interaction between light and matter at the level of single quanta enables the verification of current theories at an unprecedented level of accuracy and permits

to investigate the crossover between microscopic quantum world and our macroscopic reality.

In this thesis, superconducting circuits operated at low temperatures and microwave frequencies are investigated. A novel regime of strong light matter interaction, where the direct effect of single quanta can be observed is realized in a solid state environment [Girvin09]. The resonant and the dispersive interaction of superconducting artificial atoms embedded in high quality resonators with the electromagnetic fields is studied. The decoherence properties of such a system are assessed, with a particular emphasis on its suitability for quantum information processing. In Chapter 4, the preparation of high quality quantum states and their read-out with high fidelity is demonstrated [Bianchetti09]. In Chapter 5, the usual two level approach is generalized to three coherent states which are excited to arbitrary superpositions using optimal control techniques. The generated qutrit state fidelities are evaluated using full quantum state tomography [Bianchetti10]. Finally, in Chapter 6, the nonlinear, high measurement power system response is analyzed, demonstrating single shot read-out fidelities up to 84 %.

The experimental framework enabling such experiments was set up in the first year of the thesis and is presented in Chapter 3. Under others, the appropriate wiring of a dilution refrigerator, enabling experiments at the single photon level and gigahertz frequencies is discussed. Also the generation of high quality control signals suited for quantum information processing and the measurement of single photons with commercial microwave equipment is depicted.

1.1 Quantum information processing

The idea of computation with intrinsically quantum mechanical objects dates back to the early eighties and was formalized in terms of a quantum computer by [Bennett82, Deutsch85]. Using such a controllable quantum computer to simulate complex quantum system was first proposed by [Feynman82], while the first quantum key distribution algorithm was proposed by [Bennett84]. Further research in the nineties led to the proposal of a set of rules every physical implementation of a quantum computer should fulfill [DiVincenzo97] and the demonstration of the universality of quantum computation [Lloyd96]. Parallel to this work, the experimental realization of a Bell inequality test [Bell64] by [Aspect82], ruled out local realistic theories and demonstrated quantum entanglement [Horodecki09] experimentally for the fist time.

Diverse physical systems potentially fulfill the strict requirements of quantum computation. Using natural quantum systems, such as ions has led to early advances in the demonstration of the single building blocks of a quantum computer, see [Leibfried03, Häffner08, Duan10]

CHAPTER 1. INTRODUCTION

for a review. Ultracold neutral atoms in optical lattices are promising candidates as quantum simulators and are described in [Bloch08, Lewenstein07]. Rydberg atoms embedded in high quality cavities [Raimond01] demonstrated many of the features needed for quantum computing. Photons are best suited for quantum communication tasks, such as cryptography [Gisin02, Scarani09], but can also be used to perform computations in linear optical networks [Dell'Anno06, Kok07]. Even long time storage of photons was demonstrated at room temperature in atomic ensembles [Hammerer10].

In solid state systems, several implementations of quantum computers have been proposed. They are expected to profit from the current micro-fabrication techniques to efficiently scale up to more complex systems. Laterally defined, electrically controlled quantum dots manipulating single electrons or even single spins have been investigated [Hanson07]. Nitrogen vacancy centers in diamond demonstrated striking coherence times at room temperature [Gaebel06, Dutt07], while optically controlled, self assembled quantum dots [Skolnick04] demonstrated strong coupling [Yoshie04, Reithmaier04]. Single spins can be detected optically [Berezovsky06] and could be used as building block of future quantum computers [Cerletti05].

Also, superconducting quantum circuits [Clarke08] significantly contribute to the recent advances in quantum computing. System parameters can be designed at will and diverse properties such as coupling mechanism, energy level structure or susceptibility to different noise mechanisms can be chosen by design. Superconducting artificial atoms can be regarded as qubits which are easily initialized in the ground state, can be read-out locally, can be manipulated by a fast and universal set of gates and which can be employed for a scalable architecture in a viable way. The reduced coherence compared to atomic systems is more than offset by the flexibility of superconducting devices.

A superconducting artificial atom embedded in a cavity, proposed in [Blais04], has been implemented and is studied in this thesis. The cavity acts both as a noise suppressing environment and as a coupling device between photons and qubit. The strong coupling to photons [Wallraff04] is used to read-out single or two-qubit states [Bianchetti09, Filipp09] and as a bus to perform two-qubit operations [Majer07], while high fidelity single shot read-out was demonstrated by [Mallet09, Reed10b]. Controlled NOT gates have been demonstrated before in related systems [Yamamoto03, Plantenberg07] and where carefully assessed in [Bialczak10], while [Chow09b] benchmarked the fidelity of single qubit gates. Combining several of the mentioned methods lead to a first demonstration of a simple quantum algorithm [DiCarlo09].

Furthermore, the implementation of geometric phases [Leek07] to generate robust, high fidelity gates as well as novel cavity designs with separate storage and read-out modes [Leek10] are encouraging steps towards the goal of a viable quantum computer. A promising idea to scale up the system with a cavity grid was proposed by [Helmer09a] and even the generation of GHZ-states [Bishop09b, Helmer09b, DiCarlo10] or the implementation of a quantum simulator, for example of a spin [Neeley09] have been investigated.

1.2 Quantum optics on a chip

Early studies of the energy exchange between light and matter led to the development of quantum mechanics. The study of the interaction between single atoms and few photons trapped in a cavity led to the development of cavity quantum electrodynamics (CQED), see [Walther06] for a review and references therein. The coherent exchange of one excitation between an atom and the field in the resonant strong coupling regime is the most striking feature of such a system. The interaction is, however, going well beyond simple energy exchange and leads to dramatically different atomic decay rates and large modifications of the atomic energy spectra compared to the free space case. It enables for the realization of highly non classical Fock states and displays characteristic lasing properties. Furthermore, the atomic state can be inferred via the photons leaking out from the cavity and inversely the field state can be read-out by the escaping atoms, demonstrating correlations characteristic of entangled states.

The small mode volume realized in coplanar waveguides and the big dipole moment of superconducting artificial atoms makes circuit cavity quantum electrodynamics an ideal testbed for such phenomena. An instructive review has been published by [Girvin09]. The strong coupling regime has been demonstrated early on by measuring the vacuum Rabi mode splitting [Wallraff04], while the dispersive operation and read-out of single qubit states was demonstrated shortly later [Wallraff05]. Both the AC-Stark or light shift of the resonator [Schuster05] and the Lamb shift of the qubit [Fragner08], characteristic of the dispersive system operation where measured. Direct quantum mechanical system aspects such as the quantization of the photon number in a coherent cavity state [Schuster07b], or the generation of single photons [Houck07] and the generation of arbitrary photonic states [Hofheinz08, Hofheinz09] were demonstrated. Even lasing indications from a single artificial atoms were found in such systems [Astafiev07]. A first experiment violating the CHSH version of the Bell inequality has also been performed [Ansmann09], closing the detection loophole. Autler-Townes and Mollow

CHAPTER 1. INTRODUCTION

transitions in a strongly driven qubit were observed [Baur09], while the nonlinear response to multiple photons which could be used to realize the photon blockade effect [Birnbaum05] was spectroscopically demonstrated [Fink08]. The high power response of the first vacuum Rabi mode splitting, showing the supersplitting of each vacuum Rabi peak and the appearance of extra peaks due to the coupling to higher qubit excited states has been measured [Bishop09a]. The controlled resonant interaction of several qubits instead of several photons was also realized [Fink09], effectively implementing the Tavis-Cummings model.

In addition to the promising applications in quantum computing and classical cavity quantum electrodynamics experiments, mesoscopic quantum circuits could shed new light on the crossover between quantum and classical behavior [Fink10]. Observing the measurement back action [Clerk10] and partial collapse and revival of states [Katz06, Katz08, Jordan10] for the first time in solid state physics could also lead to new insights in modern measurement theory, experimentally addressing the quantum measurement problem [Zurek03].

Chapter 2

Circuit Quantum Electrodynamics

The fundamental components for a cavity quantum electrodynamics experiment are an isolated quantum system with an anharmonic energy spectrum and a harmonic cavity. The key parameters of such a general setup are the cavity resonance frequency ω_r, the frequency ω_0 of the lowest atomic energy level spacing and the coupling strength g [Walther06]. In a solid state implementation working at microwave frequencies, employing superconducting materials naturally provides a low dissipation environment enabling long coherence times and also suppresses low energy excitations due to the single macroscopic superconducting ground-state. Superconducting qubits [Makhlin01], which rely on the Josephson effect to generate an anharmonic energy spectrum are well suited to be embedded in transmission line cavities.

The big effective dipole moment d of superconducting qubits, combined with the enhanced zero-point electric field $\mathcal{E}_{rms}^0$ provided by the quasi one-dimensional cavity generates a large coupling coefficient $g = \mathcal{E}_{rms}^0 d/\hbar$ [Blais04]. Typical coupling strengths of $g/2\pi \sim 100$ MHz can easily exceed typical qubit energy decay rates $\gamma/2\pi \sim 100$ kHz and photon loss rates $\kappa/2\pi \sim 0.1-10$ MHz, bringing the system in the strong coupling regime [Wallraff04]. Transition frequencies of the order of 5 GHz correspond to a temperature of around 250 mK are high enough to benefit from suppressed thermal excitations when working in a cryogenic environment at ~ 20 mK. This temperature can be reached by using commercial dilution refrigerators.

In contrast to a typical atomic experiment the qubits remain inside the cavity and manipulation time is not limited by the time of flight. The solid state fabrication leads to a well defined qubit number and a fixed interaction strength which is not easy to reach with single atoms moving in the cavity [Fink09]. Furthermore, the energy levels of superconducting qubits can be tuned in a few nanoseconds, providing an unprecedented flexibility in CQED experiments [Cooper04, McDermott05].

2.1 Transmission line cavities

It is advantageous to embed a qubit in a cavity which both acts as a filter for environmental noise and as a device to achieve strong coupling to single photons. A lumped element circuit is not well suited to couple to superconducting artificial atoms, with typical transition energies in the GHz range, because of unavoidable stray inductances and capacitances in this frequency range. To realize a distributed element resonator, the coplanar waveguide (CPW) design was chosen under the many possible implementations [Pozar90] because of its favorable properties. They can easily be matched to the standard 50 Ohm of commercial microwave components for arbitrary small lateral dimensions, allowing for miniaturization and on chip integration. They also have a simple geometry and are therefore relatively easy to design and fabricate with standard lithographic methods compatible with the nano-fabrication techniques used for Josephson junctions. Superconducting CPW were used to study superconducting materials [Yoshida92, Watanabe94, Porch95], as radiation detectors [Mazin02] and for quantum information purposes [Hammer07, Gao07, O'Connell08, Göppl08, Kumar08, Barends08, Wang09b]. The small gap between the ground planes and the center conductor generates a huge vacuum field (0.2 V/m for typical devices [Blais04]) which is essential for the strong coupling to a superconducting circuit. Also high quality factors can be achieved, allowing for long coherence times. In this section the basic design properties of transmission line cavities are discussed.

One can think of a CPW as a longitudinal slice trough a coaxial cable which is intersected at two points to generate a resonant cavity, see Fig. 2.1. Following [Simons01], the impedance is

$$Z_0 = \frac{30\pi}{\sqrt{\epsilon_{eff}}} \frac{K(k_0')}{K(k_0)} \qquad (2.1)$$

for an effective dielectric constant on a thick dielectric substrate with $\epsilon_{eff} = (1+\epsilon_m)/2$. $K(x)$ is the complete elliptic integral of the first kind, $k_0 = a/b$, $k_0' = \sqrt{1+k_0^2}$. For a coplanar waveguide on sapphire ($\epsilon_m \cong 10$), with a chosen center conductor width $2a = 10$ μm one finds $b = 9.2$ μm for a 50 Ohm line. The metal thickness is not taken into account, but [Kitazawa86] predicted a change in the characteristic impedance of 2% taking into account 100 nm metallization thickness and 3% for 200 nm. Using the "TxLine" tool included in the software package [AWR-Corp.06],

CHAPTER 2. CIRCUIT QUANTUM ELECTRODYNAMICS

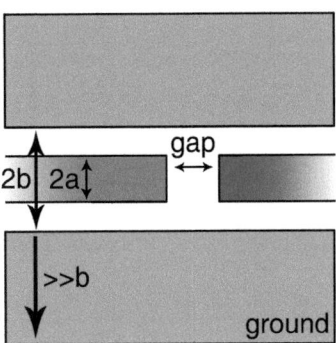

Figure 2.1 – Top view of a coplanar waveguide interrupted by a gap, effectively implementing a mirror. The ratio a/b sets the characteristic impedance Z_0 of the line, while the gap size determines the coupling between the center conductor of the feed line (in red) and the resonator (blue). Ground planes are shown in green.

which solves the problem numerically and additionally assuming a loss tangent $\tan \delta = 10^{-6}$ and a conductivity of 10^{18} S/m, $b = 9.7$ μm on sapphire and 150 nm metal thickness. We choose $b = 9.5$ μm, implying $\epsilon_{eff} = 5.5$ using the analytical approach and $\epsilon_{eff} = 5.3$ using TxLine.

For a given geometry and choice of materials, the resonance frequency of a cavity is fixed by its length. For a $\lambda/2$ resonator one can calculate the physical length which corresponds to an electrical length of 180 degrees. Using the parameters stated above, and designing 7 GHz for the first mode, one finds 9.14 mm with Eq. (2.1) and 9.31 mm with TxLine. These values do not take into account for the shift of the resonance frequency due to the coupling of the resonator to the leads [Göppl08] nor the kinetic inductance of the superconductor.

The kinetic inductance only slightly changes Z_0 if the skin depth of the superconductor stays smaller than the thickness of the metal. For our parameters, using aluminum or niobium metallizations, a small additional component to the inductance of the line has to be considered [Watanabe94]

$$Z_0 = \frac{1}{\sqrt{\epsilon_{\text{eff}}}} \left[30\pi c \mu \frac{K(k_0')}{K(k_0)} \left(\frac{K(k_0')}{4K(k_0)} + \frac{\lambda^2}{2at} c_1 \right) \right]^{\frac{1}{2}}, \qquad (2.2)$$

where c_1 is a geometrical factor, c is the speed of light in vacuum, $\mu = \mu_0 \mu_r$ the magnetic

2.1 Transmission line cavities

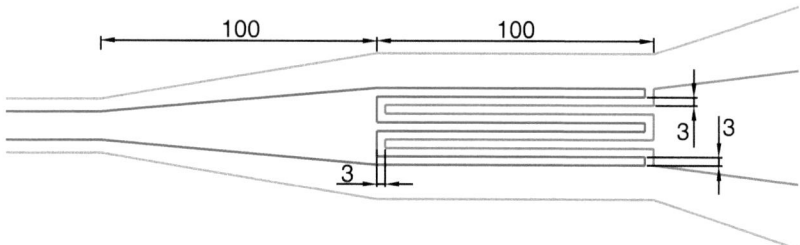

Figure 2.2 – "Finger capacitor" used to increase the coupling and design resonators with lower quality factors. All units are in μm. The aspect ratio a/b, sketched in Fig. 2.1, is kept constant while the physical dimensions are increased without changing the impedance.

permeability, λ the skin depth of the superconductor and t the thickness of the superconducting layer. For a typical resonator, the frequency shift of the resonance due to the kinetic inductance is 35 kHz (1% change in Z_0) at base temperature (20 mK), using $\mu_r = 1$, $t = 150$ nm and $\lambda = 50$ nm for aluminum. Considering other effects shifting the resonance frequency, such as fabrication imperfections or the coupling to the input/output ports [Göppl08], this is a small correction and is usually neglected.

The circuit counterparts to mirrors in optical cavities are capacitors implemented as gaps in the center conductor. The size of the two capacitances defines at which rate photons from the cavity are transmitted to the feed lines. Due to the strong interaction with the artificial atom, the photonic state is entangled with the qubit and can be used to infer its quantum state [Blais04]. The rate at which photons escape the cavity has to be bigger than the qubit decay rate to ensure an efficient state detection.

To quantitatively model the coupling, the capacitance between the feedline and resonator center conductor is calculated. Conventional optical lithography has not enough resolution to define gap capacitors bigger than about 1 fF (corresponding to a gap of 2 μm, see Fig. 2.1). The resulting cavity linewidth of 10 kHz would be too small for most read-out applications in circuit QED. To increase the coupling, a design of an interdigitated capacitor with a longer interface of the two center conductors, which we call "finger capacitor", is employed instead, see Fig. 2.2. The capacitance of such a component is calculated using a finite element simulation [Ansoft-Corp.05], finding $C = 20$ fF for the shown design.

Not only the coupling to the feed lines contributes to a finite resonance linewidths, but also

CHAPTER 2. CIRCUIT QUANTUM ELECTRODYNAMICS

internal losses have to be taken into account. It is therefore useful to define the loaded quality factor of a cavity with resonance frequency ω_r

$$Q_L := \omega_r \frac{\text{Stored energy}}{\text{Dissipated power}} = \omega_r \frac{\bar{n}\hbar\omega_r}{2P_{out} + P_{loss}}, \tag{2.3}$$

which indicates how many times a photon is reflected back and forth before it is lost. When the cavity is driven on resonance with a coherent tone, $\bar{n}$ is the average number of photons in the cavity. P_{loss} is the power dissipated in the resonator, while, in the case of symmetrical coupling capacitances, P_{out} indicates the power coupled to the input and output ports. The loaded quality factor is therefore usually expressed as the reciprocal sum of the internal quality factor Q_{int} defined by internal losses and the external quality factor Q_{ext} coming from photons leaking to the leads

$$Q_{int} := \omega \frac{\bar{n}\hbar\omega_r}{P_{loss}}, \tag{2.4a}$$

$$Q_{ext} := \omega \frac{\bar{n}\hbar\omega_r}{2P_{out}}, \tag{2.4b}$$

$$\frac{1}{Q_L} = \frac{1}{Q_{int}} + \frac{1}{Q_{ext}}. \tag{2.4c}$$

Using a simple RLC model, a resonator connected on both sides to identical input and output capacitors C, has a loaded quality factor of [Pozar90, Schuster07a, Göppl08]

$$Q_{symm} = \frac{l\pi}{4Z_0}\left(\frac{1}{R_L C^2 (l\omega_r)^2} + R_L\right), \tag{2.5}$$

for the harmonic mode with index l and a load resistance R_L connected to each lead. The analytic RLC approximation is very good close to resonance (less than 0.5% deviations from a model implying $ABCD$ matrices at 10 MHz detuning for a resonator with Q_L =1000) but shows strong deviations between two modes where a full analysis has to be performed.

A charge qubit placed at a voltage node of the field is coupled via an effective dipole interaction [Blais04, Girvin09]. To calculate such a coupling, the root mean square zero-point electric field $\mathcal{E}_{rms}^0$ at the qubit position must be inferred. [Blais04, *et al.*] describe how to

11

canonically quantize a transmission line cavity, finding

$$V_{rms}^0 = \sqrt{\frac{\hbar \omega_r}{C_L L}}, \qquad (2.6)$$

where $C_L = (\epsilon_{eff}\pi)/(2L\omega_r Z_0)$ is the capacitance per unit length of the resonator and L is its physical length. At the center of the cavity, a field antinode is present for the first harmonic $l = 2$, which with a fundamental resonant frequency of 5 GHz, generates a voltage of $V_{rms}^0 \approx 1~\mu$V. This voltage drop between the center conductor and the ground plane, separated by $b - a = 4.5~\mu$m implies an electric field $\mathcal{E}_{rms}^0 \approx 0.2$ V/m, which is generated by vacuum fluctuations alone.

2.2 The Cooper-pair box

A qubit is usually defined as a quantum two-level system with well defined, long lived states and well known coupling characteristics [Nielsen00]. Engineering such a system with the conventional components of superconducting electronic circuits, containing only capacitors (C), inductors (L) and resistors (R) is, however, not possible. A dissipation free LC circuit only realizes an harmonic oscillator with characteristic Hamiltonian

$$H_{LC} = \frac{\hat{\phi}^2}{2L} + \frac{\hat{q}^2}{2C} = \hbar\omega_r(\hat{a}^\dagger \hat{a} + \frac{1}{2}). \qquad (2.7)$$

$\hat{\phi}$ and $\hat{q}$ denote the canonical conjugate variables for flux and charge respectively and $\hat{a}^{(\dagger)}$ denote the bosonic ladder operators, annihilating (creating) a single excitation/photon. The equally spaced energy levels impede an individual addressing and the system cannot be restricted to two-dimensional subspace.

A nonlinear element is needed to realize an anharmonic spectrum, but canonical components such as diodes or transistors are intrinsically dissipative and therefore not suited to maintain coherence for long times. The Josephson current flowing between two weakly linked superconducting materials [Tinkham96] through a thin tunnel barrier provides, however, such a non-dissipative non-linear response. The voltage V and current I are governed by the basic

CHAPTER 2. CIRCUIT QUANTUM ELECTRODYNAMICS

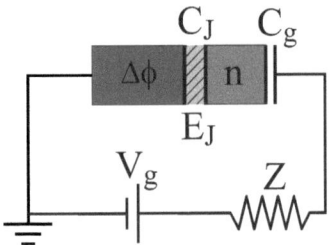

Figure 2.3 – Sketch of a Cooper-pair box. A superconducting island (red), containing n Cooper-pairs is separated by a tunnel barrier (orange) from a bulk superconductor (blue), forming a Josephson-junction with Josephson energy E_J and capacitance C_J. An external gate voltage V_g is applied through the impedance Z to the capacitance C_g.

equations

$$V(t) = \Phi_0 \frac{\partial [\Delta\phi(t)]}{\partial t}, \quad (2.8)$$

$$I(t) = I_c \sin [\Delta\phi(t)], \quad (2.9)$$

where $\Delta\phi$ is the difference in the phase factor of the Ginzburg-Landau complex order parameter of the two superconductors. Below the maximal current I_c no voltage drop across the barrier appears and therefore no power is dissipated.

A Josephson-junction can be voltage or current biased, leading to two broad classes of devices named after the degree of freedom which characterizes the basis states best. The first is defined by a small capacitance of the Josephson junction leading to a well defined number of Cooper-pairs on a superconducting island while the latter is best described by the superconducting phase difference across the junction.

A voltage biased Josephson-junction is shown in Fig. 2.3, where a small superconducting box (red) is separated by the junction (orange) on one side and a small capacitance C_g on the other from the bulk of the superconductor, realizing the so called Cooper-pair box (CPB) qubit. The system is coupled to an external voltage source V_g via an environmental impedance Z. The relevant energies are the single electron charging energy E_C, needed to add a Cooper-pair to

2.2 The Cooper-pair box

the box and the Josephson energy

$$E_C = \frac{e^2}{2(C_g + C_J)}, \qquad (2.10)$$

$$E_J = \frac{I_c \Phi_0}{2\pi}, \qquad (2.11)$$

where $\Phi_0 = h/2e$ is the magnetic flux quantum. The system is described by the Hamiltonian [Makhlin01]

$$\begin{aligned} H &= 4E_C(\hat{N} - N_g)^2 - E_J \cos\left(\widehat{\Delta \phi}\right) \\ &\approx \sum_N \left[4E_C\left(N - N_g\right)|N\rangle\langle N| - \frac{E_J}{2}\left(|N+1\rangle\langle N| + |N\rangle\langle N+1|\right) \right]. \end{aligned} \qquad (2.12)$$

Here $\hat{N}$ is the number operator of Cooper-pairs on the island, while $\widehat{\Delta \phi}$ denotes the phase difference of the superconducting order parameter, while $N_g = C_g V_g / 2e$ is the dimensionless gate charge acting as a control parameter. In the charge regime ($4E_C \gg E_J$), the approximation of Eq. (2.12) holds and the charge states $|N\rangle$ form a convenient basis. In the opposite limit, a simple phase qubit in a superconducting ring geometry is realized, setting $V_g = Z = C_g = 0$ and choosing a much bigger C_J, such that E_J dominates over E_c. In this case the phase (or current) is well suited as basis states and the environment couples trough the flux Φ enclosed by the ring [Makhlin01].

Considering only the two lowest charge states $|0\rangle$ and $|1\rangle$ and assuming $N_g \in [0,1]$, the Hamiltonian reduces to a 2x2 matrix [Blais04]

$$H = -\frac{E_{el}}{2}\hat{\sigma}_z - \frac{E_J}{2}\hat{\sigma}_x, \qquad (2.13)$$

with $E_{el} = 4E_C(1 - 2N_g)$ and $\hat{\sigma}_i$ the pauli matrixes. This realizes an effective 2-level system, with effective fields in the x and z directions.

E_{el} can be tuned with the control parameter N_g, while E_J is fixed by fabrication. This can be changed replacing the single Josephson junction by a parallel pair of junctions, each with energy $E_J/2$, which form a loop, enclosing the flux Φ. The superconducting quantum

CHAPTER 2. CIRCUIT QUANTUM ELECTRODYNAMICS

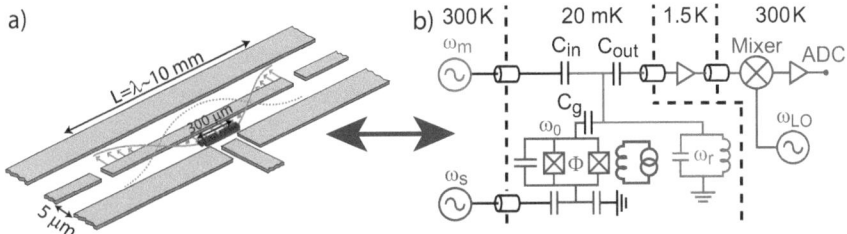

Figure 2.4 – a) Schematic of a superconducting qubit embedded in a transmission line resonator, realizing a cavity QED setup. The qubit is placed at a voltage antinode of the cavity, coupling capacitively to the field. Additionally, a capacitively coupled 50 Ohm line can be used to directly manipulate the qubit. b) Circuit diagram of the experimental setup. A harmonic oscillator modeled as an LC circuit with resonance frequency ω_r is coupled to a transmon-type qubit through the effective coupling capacitance C_g. The qubit transition frequency ω_0 is controlled by an externally applied magnetic flux Φ. The qubit state is coherently manipulated by a pulsed microwave source at the frequency ω_s. The resonator is probed by a signal applied to the input capacitor C_{in} at the frequency ω_m. The transmitted signal is amplified and down-converted by mixing with a local oscillator at frequency ω_{LO} and then digitized using an analog to digital converter (ADC).

interference device (SQUID) formed in this way has a modified potential energy of the form

$$E_J \cos\left(\frac{\pi \Phi}{\Phi_0}\right) \cos\left(\hat{\Delta\phi}\right), \tag{2.14}$$

so that both effective fields can be controlled externally, inducing transitions or changing the effective transition frequencies.

2.3 Reaching the strong coupling regime of circuit QED

As discussed in Sec. 2.1, even the usually tiny vacuum fluctuations generate large voltages between the center conductor and the ground planes of the resonator. The description of a charge qubit placed in the resonator gap must therefore take into account for this quantum effect by adding the term $V_{rms}^0 \left(\hat{a}^\dagger + \hat{a}\right)$. Following [Blais04], the Hamiltonian of the coupled system, sketched in Fig. 2.4, restricted to the two-dimensional qubit subspace, as in Eq. (2.13)

2.3 Reaching the strong coupling regime of circuit QED

reads

$$H = \hbar\omega_r \left(\hat{a}^\dagger \hat{a} + \frac{1}{2}\right) + \frac{\hbar\omega_0}{2}\hat{\sigma}_z - e\frac{C_g}{C_J + C_g}\sqrt{\frac{\hbar\omega_r}{C_L L}} \left(\hat{a}^\dagger + \hat{a}\right) \times$$
$$\times \left[1 - 2N_g - \cos(\theta)\hat{\sigma}_z + \sin(\theta)\hat{\sigma}_x\right], \tag{2.15}$$

where $\omega_0 = \sqrt{E_J^2 + [4E_C(1-2N_g)]^2}/\hbar$ is the energy splitting of the qubit and $\theta = \arctan[E_J/4E_C(1-2N_g)]$ the mixing angle. At the charge degeneracy point $N_g = 1/2$, $\theta = \pi/2$ and $\omega_0 = E_J/\hbar$, Eq. (2.15) reduces to the well known Jaynes-Cummings Hamiltonian [Jaynes63]

$$H_{JC} = \hbar\omega_r \left(\hat{a}^\dagger \hat{a} + \frac{1}{2}\right) + \frac{\hbar\omega_0}{2}\hat{\sigma}_z + \hbar g \left(\hat{a}^\dagger \hat{\sigma}_- + \hat{\sigma}_+ \hat{a}\right), \tag{2.16}$$

where $\hat{\sigma}_{+/-} = (\hat{\sigma}_x \pm i\hat{\sigma}_y)/2$ are the raising and lowering operator for the qubit. This Hamiltonian is obtained performing the rotating wave approximation (RWA), effectively neglecting the terms $\hat{a}^\dagger \hat{\sigma}_+$ and $\hat{a}\hat{\sigma}_-$ which do not conserve the number of excitations in the system. If one transforms into the frame rotating at the frequency ω_r for the cavity and ω_0 for the qubit, these terms oscillate at the frequency $\omega_r + \omega_0$ which is usually very high and can therefore be dropped. The coupling strength

$$g = \frac{C_g e}{\hbar(C_g + C_J)}\sqrt{\frac{\hbar\omega_r}{C_L L}}, \tag{2.17}$$

can exceed 100 MHz with realistic coupling parameters C_i and is therefore much bigger than any decay rate in the system. If damping is neglected, the exact diagonalization leads to the eigenstates [Yamamoto99]

$$|n, +\rangle = \cos\theta_n |e, n-1\rangle + \sin\theta_n |g, n\rangle, \tag{2.18}$$
$$|n, -\rangle = -\sin\theta_n |e, n-1\rangle + \cos\theta_n |g, n\rangle, \tag{2.19}$$

where $n > 0$ is the number of photons. The corresponding eigenenergies are

$$E_{\pm,n} = n\hbar\omega_r \pm \frac{\hbar}{2}\sqrt{4g^2 n + \Delta^2}, \tag{2.20}$$

CHAPTER 2. CIRCUIT QUANTUM ELECTRODYNAMICS

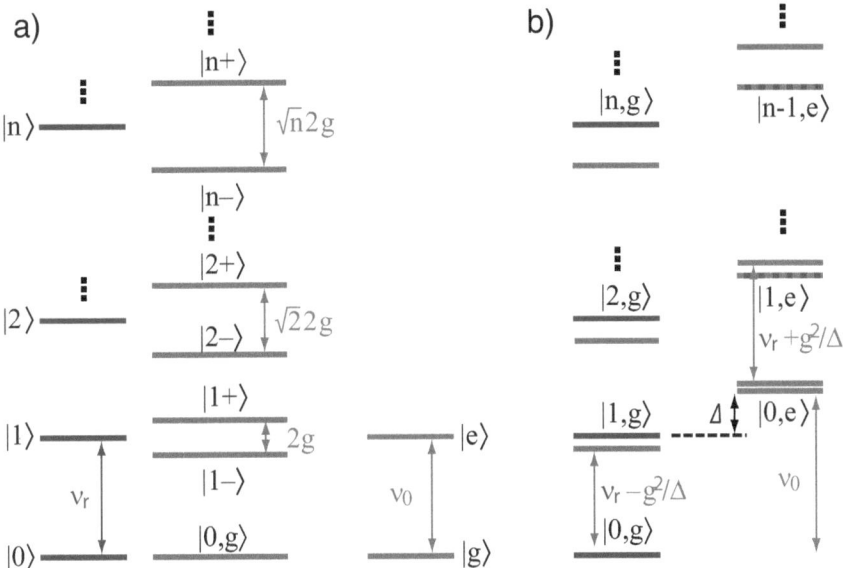

Figure 2.5 – a) Energy spectrum of the uncoupled cavity (left, blue) with the characteristic equidistant levels and qubit (right, red). The dressed qubit-photon states are shown in the center in the case of zero detuning ($\Delta = 0$). The degeneracy of the states is lifted by the strong coupling g, forming symmetric and antisymmetric doublets consisting of n photons and a qubit excitation, having spacing $2g\sqrt{n}$. b) Energy spectrum in the dispersive regime ($\Delta \gg g$). The energies are conditionally shifted by the qubit state.

and the ground state $|0, g\rangle$ has an energy $E_{g,0} = -\hbar\Delta/2$. $\Delta = \omega_0 - \omega_r$ is the atom-cavity detuning and

$$\theta_n = \arctan\left(\frac{\Delta - \sqrt{\Delta^2 + 4g^2 n}}{2g\sqrt{n}}\right). \tag{2.21}$$

In the case of zero detuning ($\Delta = 0$), sketched in Fig. 2.5a, the excited states consist of maximally entangled atom-field states [Rempe87] which therefore decay with the combined rate $(\kappa + \gamma)/2$. The formed doublets have Rabi splittings which scale with the square root of the number of photons [Brune96, Fink08]. This effect cannot be explained by any purely classical theory (contrariwise to the Rabi splittings explained classically as the normal mode splitting of

17

two coupled linear oscillators), demonstrating the purely quantum nature of the system which should enable for novel effects such as the photon blockade [Birnbaum05]. Furthermore, this can be used to generate Fock states by preparing the qubit in the excited state while the cavity is empty and strongly off-resonant and then tune the qubit to resonance for the time needed to swap the excitation to the cavity and at the end detune the qubit again [Hofheinz08, Bozyigit10].

2.4 Dispersive regime

The operation of the qubit at large detunings from the cavity resonance ($\lambda = g/\Delta \ll 1$) realizes a different regime where the combined eigenstates are basically the tensor product of the uncoupled qubit and resonator states and only their energies are shifted.

To get insight in the dispersive regime of the Hamiltonian (2.16), it is useful to introduce the unitary transformation [Blais04, Boissonneault09]

$$U = \exp\left[\frac{\arctan\left(2\lambda\sqrt{\hat{N}_e}\right)}{2\sqrt{\hat{N}_e}}(\hat{a}\hat{\sigma}_+ - \hat{a}^\dagger\hat{\sigma}_-)\right] \approx \exp\left[\lambda(\hat{a}\hat{\sigma}_+ - \hat{a}^\dagger\hat{\sigma}_-)\right], \qquad (2.22)$$

where $\hat{N}_e = \hat{a}^\dagger\hat{a} + (\hat{\sigma}_z + 1)/2$ is the number operator for excitations in the system, commuting with the operators $(\hat{a}\hat{\sigma}_+ \pm \hat{a}^\dagger\hat{\sigma}_-)$ and H_{JC}, so that

$$\begin{aligned}H_{JC}^D &= \hbar\omega_r\hat{a}^\dagger\hat{a} + \hbar\left[\omega_0 - \Delta\left(1 - \sqrt{1 + 4\lambda^2\hat{N}_e}\right)\right]\frac{\hat{\sigma}_z}{2} & (2.23)\\ &\approx \hbar\omega_r\left(\hat{a}^\dagger\hat{a} + \frac{1}{2}\right) + \hbar\left(\omega_0 + \frac{g^2}{\Delta}\right)\frac{\hat{\sigma}_z}{2} + 2\hbar\frac{g^2}{\Delta}\hat{a}^\dagger\hat{a}\frac{\hat{\sigma}_z}{2}, & (2.24)\end{aligned}$$

can be calculated exactly. Equation (2.24) is a first order approximation valid only for $4\lambda^2\hat{N}_e \ll 1$. This naturally defines the critical photon number $n_{crit} = 1/(4\lambda^2)$ above which the dispersive approximation breaks down. In the regime of large detuning and small photon numbers, however, this Hamiltonian is a very good approximation. The Lamb shift of $g^2/(2\Delta)$ and the AC-Stark shift of g^2/Δ per photon both shifting the qubit become apparent from this expression. The AC-Stark shift can alternatively be interpreted as a dispersive shift of the cavity transmission frequency dependent on the qubit state, enabling for a qubit state determination by the monitoring of the cavity field, as described in Ch. 4.

Coupling a transmission line resonator to a qubit is not restricted to charge qubits. In phase qubits, however, high single shot measurement fidelities (90%) [Lucero08] are already available. Thus, the cavity is not used as a measurement device for the qubit, but as a resource which can be manipulated by the qubit to demonstrate, for example, the creation of Fock states in an harmonic resonator [Hofheinz08, Hofheinz09] and their decay [Wang08, Wang09a]. In such a hybrid system the qubit with his strong interaction performs quantum computations and the photonic mode can be used as a resource with long coherence time to store quantum information [Leek10]. The generated photons are also suited to carry quantum information and perform communication tasks between separated systems.

The eigenstates (2.18) can also be approximated to first order in λ

$$|-,n\rangle \approx |g,n\rangle + \sqrt{n}\lambda|e,n-1\rangle, \qquad (2.25)$$

$$|+,n-1\rangle \approx |e,n-1\rangle - \sqrt{n}\lambda|g,n\rangle, \qquad (2.26)$$

making manifest that the states are only slightly mixed by the small parameter λ. Using Fermi's golden rule, it can be shown that this small photonic component opens an additional spontaneous decay channel for the qubit excited state, called Purcell effect [Walls94] which has been observed with Rydberg atoms [Goy83] and in circuit QED [Houck07, Houck08]

$$\gamma_\kappa = \kappa \frac{g^2}{\Delta^2}. \qquad (2.27)$$

2.5 Transmon artificial atom

Low frequency noise coupling to the qubits leads to dephasing. This is one of the challenges which still need to be solved to realize a quantum computer. As seen in Sec. 2.2, the transition energy of CPB type qubits is tuned via an externally applied gate voltage V_g which is affected by $1/f$ type noise limiting the phase coherence time T_2 of the two level systems [Abragam61, Ithier05]. Charge noise in such systems is the dominant noise contribution, strongly limiting the coherence times. One possible solution to suppress this problem is to operate the qubit at its "sweet-spot", where the charge dispersion is independent of the applied voltage to first order [Vion02]. A second approach is to decrease the charge dispersion by increasing the E_J/E_C ratio and therefore operate the qubit in a different regime [Koch07]. The charge dispersion can be decreased exponentially while the anharmonicity of the energy ladder, needed to operate the qubit selectively, only decreases with a weak power law. Since the

2.5 Transmon artificial atom

qubits are capacitively coupled to the resonator, one would intuitively expect also a reduced cavity coupling. Surprisingly, however, the coupling strength is not affected, in the contrary it can even be increased. Designing the charge dispersion smaller than the energy relaxation time T_1 also makes DC voltage lines superfluous, enabling for a simpler experimental setup, see Sec. 3.1.

To increase the E_J/E_C ratio while keeping the transition frequency in the GHz range, $E_C = e^2/(2C_\Sigma)$ can be decreased by increasing the effective $C_\Sigma = C_J + C_g + C_s$, via an additional big shunt capacitance C_s across the Josephson junction [Koch07]. Formally the same Hamiltonian as in Eq. (2.12) is realized. The new qubit design is named transmission-line shunted plasma oscillation qubit or "transmon". In a regime where $E_J \gg E_C$, the phase basis is best suited to diagonalize the Hamiltonian which can be solved analytically and exactly using Mathieu functions. For $E_J/E_C \gtrsim 10$ the eigenenergies can be approximated by [Koch07]

$$E_m \approx -E_J + \sqrt{8E_C E_J}\left(m + \frac{1}{2}\right) - \frac{E_C}{12}\left(6m^2 + 6m + 3\right) \qquad (2.28)$$

resulting in an anharmonicity $\alpha = E_1 - E_0 \approx -E_C$ and a charge dispersion

$$\begin{aligned}\varepsilon_m &= E_m(n_g = 1/2) - E_m(n_g = 0) \\ &\approx (-1^m) E_C \frac{2^{4m+5}}{m!}\sqrt{\frac{2}{\pi}}\left(\frac{E_J}{2E_C}\right)^{\frac{m}{2}+\frac{3}{4}} e^{-\sqrt{8E_J/E_C}}.\end{aligned} \qquad (2.29)$$

For typical parameters $E_{01}/\hbar = 6$ GHz, $E_C/\hbar = 300$ MHz, the system anharmonicity enables for fast operations if the pulse shapes are optimized [Motzoi09] and the charge dispersions $\varepsilon_0 = 10$ kHz, $\varepsilon_1 = 300$ kHz are small.

In analogy to Eq. (2.13), the transmon embedded in a cavity can be described by the generalized Jaynes-Cummings Hamiltonian [Koch07],

$$H_{JC} = \hbar\omega_r \hat{a}^\dagger \hat{a} + \hbar \sum_{i=0}^{M} \omega_i |i\rangle\langle i| + \hbar \sum_{ij=0}^{M-1} g_{ij}\left(|i\rangle\langle j|\hat{a} + \hat{a}^\dagger |j\rangle\langle i|\right), \qquad (2.30)$$

where $|i\rangle$ describes the eigenstate with energy E_i and the coupling between state $|i\rangle$ and $|j\rangle$ is given by

$$\hbar g_{ij} = 2\beta e V_{rms}^0 \langle i|\hat{N}|j\rangle. \qquad (2.31)$$

CHAPTER 2. CIRCUIT QUANTUM ELECTRODYNAMICS

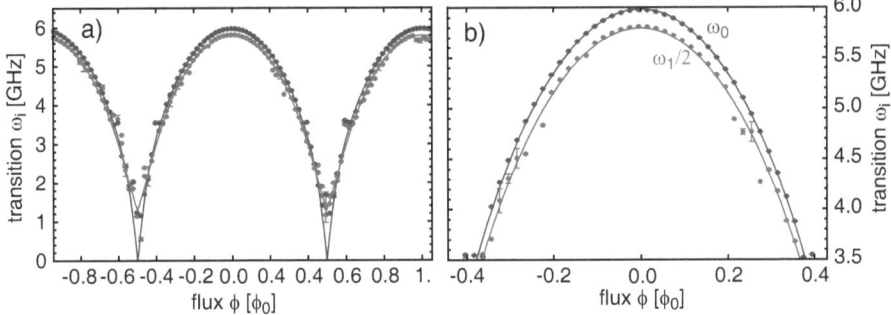

Figure 2.6 – a) Transmon spectrum as a function of the applied magnetic field measured spectroscopically. The blue line shows the ω_0 transition, while the red one $\omega_1/2$. b) Magnification of panel a, showing the region around the maximal transition frequency.

Again, for this expression, $\hat{N}$ is the number operator for the Cooper pairs, while $\beta = C_g/C_\Sigma$ is the capacitance ratio. In the limit of small E_C/E_J, the matrix elements $\langle i|\hat{N}|j\rangle$ for non nearest neighbors vanish and

$$\langle j+1|\hat{N}|j\rangle \approx \sqrt{\frac{j+1}{2}}\left(\frac{E_J}{8E_C}\right)^{1/4}. \qquad (2.32)$$

Even if the DC voltage coupling is exponentially suppressed, the AC response increases as $(E_J/E_C)^{1/4}$ and the transmon can be strongly coupled to the cavity.

An example of a transmon spectrum versus applied magnetic flux in the superconducting SQUID loop is shown in Fig. 2.6a and is magnified in panel b. The transition frequencies are measured spectroscopically [Schreier08], while the theoretical predictions are calculated evaluating numerically the Mathieu functions and taking into account for a small asymmetry d between the two Josephson junctions [Koch07]. The blue trace depicts the frequency of the first transmon transition ω_0, showing the flux periodicity expected from Eq. (2.14). The red line shows the second transmon transition frequency ω_1 which is driven with two photons and therefore appears at the frequencies $\omega_1/2$ in the plot. The charging energy $E_C/2\pi = 326$ MHz and the maximal Josephson energy $E_J/2\pi = 15.4$ GHz result from a fit to the measured data, with a junction asymmetry $d = 8\%$.

2.6 Dispersive transmon regime

The dispersive approximation discussed in Sec. 2.4, applied to the transmon multi level structure results in [Boissonneault10]

$$H_{JC}^D/\hbar = \omega_r \hat{a}^\dagger \hat{a} + \sum_{i=0}^{M-1} \omega_i |i\rangle\langle i| + \sum_{i=0}^{M-1} L_i |i\rangle\langle i| + \sum_{i=0}^{M-1} S_i |i\rangle\langle i| \hat{a}^\dagger \hat{a} + \sum_{i=0}^{M-1} K_i |i\rangle\langle i| (\hat{a}^\dagger \hat{a})^2, \quad (2.33)$$

where L_i, S_i, K_i are the Lamb-shift, Stark-shift and self-Kerr effect coefficients respectively of level i, which to third order in λ_i compute to

$$\begin{aligned}
L_i &= \chi_{i-1}(1 - \lambda_{i-1}^2) - \frac{1}{2}(3\chi_{i-2}\lambda_{i-1}^2 + \chi_{i-1}\lambda_{i-1}^2) \quad (2.34)\\
&\approx \chi_{i-1},\\
S_i &= \chi_{i-1}(1 - \lambda_i^2 - 2\lambda_{i-1}^2 - \frac{3}{4}\lambda_{i-2}^2) + \frac{9}{4}\chi_{i-2}\lambda_{i-1}^2\\
&\quad -\chi_i(1 - \lambda_{i-1}^2 - \frac{1}{4}\lambda_{i+1}^2) \approx \chi_{i-1} - \chi_i, \quad (2.35)\\
K_i &= \frac{3}{4}\chi_{i-2}\lambda_{i-1}^2 - \chi_{i-1}(\frac{1}{4}\lambda_{i-2}^2 + \lambda_i^2 + \lambda_{i-1}^2) - \frac{3}{4}\chi_{i+1}\lambda_{i-1}^2\\
&\quad +\chi_i(\frac{1}{4}\lambda_{i+1}^2 + \lambda_i^2 + \lambda_{i-1}^2) \approx 0, \quad (2.36)
\end{aligned}$$

with the generalized detunings $\Delta_i = (\omega_{i+1} - \omega_i) - \omega_r$, coupling coefficients $g_i \equiv g_{i,i+1}$, dispersive shifts $\chi_i = g_i^2/\Delta_i$ and small parameters $\lambda_i = -g_i/\Delta_i$, where $\chi_i = \lambda_i = 0 \; \forall \; i \notin [0, M-2]$.

An example of the dispersively shifted cavity resonance frequency ω_r is showed in Fig. 2.7, measured with the same sample investigated in Fig. 2.6. The transmon energy level structure dispersively "pushes" the cavity resonance frequency to higher frequencies in function of the applied magnetic flux, showing a larger shift when the first excited state is nearer resonance, around $\phi = 0$. The data is fitted to Eq. (2.33), finding the coupling coefficient $g_0/2\pi = 111$ MHz and the cavity resonance frequency $\omega_r/2\pi = 6.9468$ GHz.

Considering again only the two lowest states $|g\rangle$ and $|e\rangle$, the dispersive Jaynes-Cummings Hamiltonian reduces to a 2x2 matrix, analog to Eq. (2.24)

$$H_{eff} = \hbar \omega_0' \frac{\hat{\sigma}_z}{2} + (\hbar \omega_r' + \hbar \chi' \hat{\sigma}_z) \hat{a}^\dagger \hat{a}, \quad (2.37)$$

CHAPTER 2. CIRCUIT QUANTUM ELECTRODYNAMICS

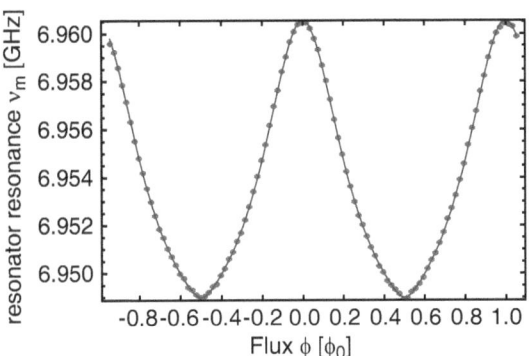

Figure 2.7 – Measured cavity resonance frequency ω_m versus magnetic flux. The blue line, fitted to the data, evaluates S_0 to calculate the pulled cavity frequency.

but with renormalized transition frequencies $\omega'_r = \omega_r - \chi_1/2$, $\omega'_0 = \omega_1 - \omega_0 + \chi_0$ and

$$\chi' = \chi_0 - \chi_1/2 \approx -\left(\beta e V_{rms}^0\right)^2 \left(\frac{E_J}{2E_C}\right)^{1/2} \frac{E_C}{\hbar^2 \Delta_0 \left(\hbar \Delta_0 - E_C\right)}. \qquad (2.38)$$

The effective dispersive shift can therefore change sign and even diverge when the second excited state gets resonant with the cavity, where the dispersive approximation breaks down.

Chapter 3

Experimental Setup

Connecting the quantum world, where single quanta produce observable effects, to the classical environment in the laboratory, where a single mobile phone can generate 10^{18} photons per second, is a challenging task. Classical control signals must be generated and "fed" to the sample space minimizing different noise sources while weak fields from the experiment have to be strongly amplified to be detected (up to a factor 10^9 in power). Additionally, the sample has to be kept at millikelvin temperatures to avoid incoherent populations caused by thermal excitations. Not only the standard phonon temperature has to be kept low but also the black body radiation temperature must fulfill the strict requirement $k_B T \ll \hbar\omega$. Moreover, the long time stability of the environmental conditions, such as the magnetic field, must satisfy stringent criteria to avoid state preparation imperfections and dephasing.

In this chapter all these aspects will be addressed, with a particular emphasis on the implementation chosen in the Quantum Device Lab at ETH.

3.1 Cryogenic wiring

Accomplishing different tasks such as detecting single photons and shielding the experiments from room temperature radiation, requires different kinds of cabling connecting the experiment to the laboratory equipment. The entire wiring, however, must not transport more heat than the cryostat can handle on every single cooling stage. For a presentation of the operation of a dilution refrigerator and a discussion of low temperature heat flows one can consult [Pobell06].

3.1 Cryogenic wiring

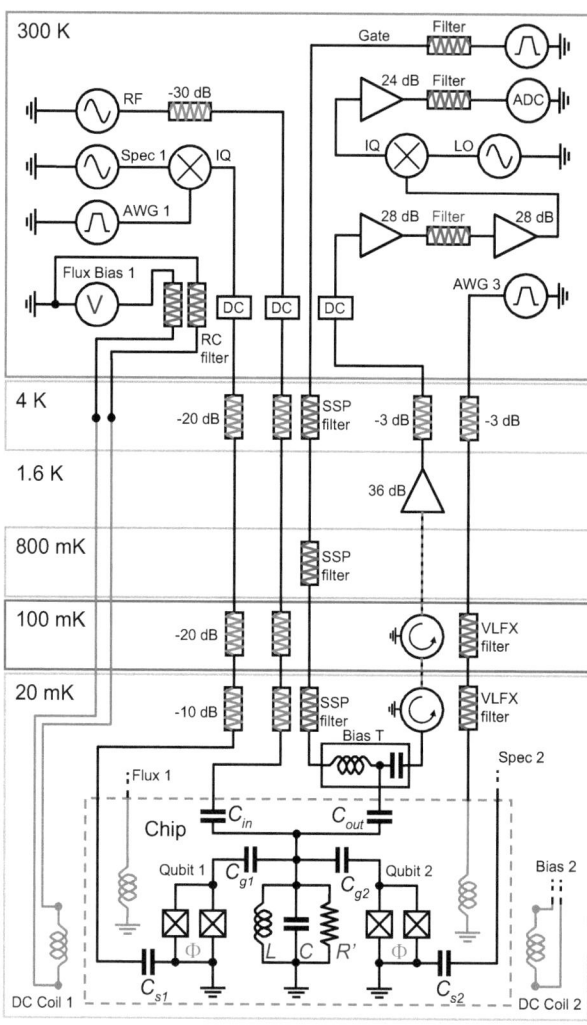

Figure 3.1 – Schematics of the measurement setup used in CQED.

CHAPTER 3. EXPERIMENTAL SETUP

Conventional and pulse tube cooler based refrigerators such as the Vericold DR200 have typical cooling powers of less than 1 W at the 4 Kelvin stage. The installation of a single semirigid tin plated copper coaxial cable (UT85 TP) from room temperature to this stage would already transfer 0.6 W of heat, considerably warming up the cryostat. For this reason, all coaxial cables connecting different temperature stages are made of stainless steel (UT85 SS) which has much lower thermal conductivity at the price of greatly decreased electrical conductivity (copper cables have an attenuation of 2.2 dB per meter at 10 GHz, while stainless steel attenuate 13.3 dB), for a detailed discussion of the involved heat flow one can consult the Appendix A.1. The only exception is the "measurement" line, carrying the field from the sample to the first cold amplifier, where losses must be avoided to maximize the signal to noise ratio of the read-out. For this reason a special cable with silver plated center conductor (4.8 dB attenuation per meter at 10 GHz) or a superconducting niobium-titanium cable is installed.

Higher temperature stages of the cryostat have more cooling power than plates at lower temperature. It is therefore essential to thermalize the cabling at each stage. To ensure a complete thermalization of the center conductor a dissipative element such as an attenuator or a component like a circulator must be employed.

The thermalization of the center conductor at different temperatures also provides the opportunity to dissipate the incident thermal radiation from a 50 Ohm load at higher temperature. The black body radiation emitted at 300 K in the microwave frequency range up to 40 GHz, is about 80 times more intense than the radiation emitted at 4 K. Therefore a 20 dB attenuator is mounted at the 4 K plate in the "control" lines (connected to the resonator input capacitor C_{in} and to the qubit via C_s), as shown in Fig. 3.1. These lines are connected to the input port of the resonator and to the qubit charge lines and are used to populate the cavity and drive direct qubit transitions respectively. The remaining radiation, corresponding to about 40 photons at 5 GHz is then attenuated at the 100 mK and base plate. The one-dimensional black body radiation is further discussed in Appendix A.2.

Getting rid of the incoming black body radiation with attenuators is not practical for the output port of the cavity C_{out}. The weak signal with an average of less then 1 photon would as well be attenuated, compromising the signal to noise ratio (SNR). For this reason, circulators are employed and the incoming radiation is routed to a 50 Ohm termination where it is absorbed, while the signal is transmitted to a first low-noise cold amplifier. Low temperature circulators from Pamtech and Raditek have isolations of at least 18 dB in the operating frequency range and have been successfully tested at 4 K, see Appendix A.3. The noise properties of the wiring

3.1 Cryogenic wiring

T	Component	Att. [dB]	Component	Att. [dB]
RT	Source out	+12/-90	ADC	
RT	UT85 TP	-1.2	UT85 TP	-1.9
RT	Attenuator	-10	Amplifier	+24
RT	Splitter	-3	UT85 TP	-0.2
RT	Attenuator	-30	IQ Downconverter	8
RT	Splitter	-3	UT85 TP	-0.9
RT	DC block	-0.5	Amplifier	+36
RT			Attenuator	-3
RT			Filter	-0.1
RT			Amplifier	+28
RT			DC block	-0.5
RT	UT85 TP	-4.2	UT85 TP	-2
-	UT85 SS	-4	UT85 SS	-4
4K	Attenuator	-20	Attenuator	-3
-	UT85 SS	-12.3	UT85 SS	-0.7
1.6K			Amplifier	+35
-			UT85 SS	-3.6
100mK			Circulator	-0.3
-			UT85 SS	-1.2
Base	Attenuator	-20	Circulator	-0.3
Base	Bias-tee	-1.2	Bias-tee	-1
Base	UT85 TP	-0.5	UT85 TP	-0.5
Base	Sample-holder	-1	Sample-holder	-1
	Total	**-110.9**	**Total**	**+90.8**

Table 3.1 – Detailed list of the employed components and corresponding attenuation factors for the main cabling used for the input and output resonator lines for the original wet Oxford fridge.

and the cold amplifier mounted on the 1 K pot in the traditional, liquid helium precooled fridge and on the cold plate of the pulse tube cooler in the DR200 fridge is analyzed in Sec. 3.2. A complete list of the components installed originally in the input and output cavity lines is shown in Tab. 3.1, while a photograph of the wiring implemented in the Vericold DR200 is shown in Fig. 3.2.

Early Cooper-pair box samples had the bias voltage as additional control parameter. The attenuation built-in in the described microwave lines would, however, not allow for enough tunability and the currents flowing in the attenuators would warm up the system. Therefore a separate line carrying the DC voltage is implemented and connected to the sample with a bias-tee from Anritsu (K250 with an RF bandwidth of 0.1-40 GHz and a DC bandwidth up to 10 MHz). The K251 model, which has a 3 dB point on the DC side of 23 kHz and a good transmission on the RF side starting at 50 kHz is used to apply quasi DC signals on the RF side. The noise is filtered with a modified stainless steel low-pass powder filter (SSPF) similar to the one described by [Lukashenko08]. High frequencies are strongly damped by the skin effect while the low frequency smooth cutoff at around 10 MHz is realized using a built-in lumped element RC-filter, see Appendix A.4.

The on-chip flux-lines, designed to quickly and locally shift the qubit transition frequency, require higher bandwidths (> 100 MHz to enable for 10 ns long flux pulses), smoother cutoffs (to avoid pulse degradation), and lower DC-damping to not warm up the sample due to the current flow. For this reason two commercial Mini-Circuits VLFX-300 low pass filters, with a 3dB point at around 450 MHz where tested at 4 K and installed in series in the cryostat. A constant current of 1 mA could be safely passed trough the chip without warming up the baseplate significantly and 10 ns long pulses could be successfully implemented [Bozyigit10].

The DC wiring of the external coils, used to tune the qubit transition frequencies is also shown in Fig. 3.1, for a full description of the filtering please refer to Sec. 3.5, while the room temperature signal modulation and demodulation is described in Sec. 3.3 and 3.4 respectively. The DC-blocks on the RF lines avoid ground loops, through which induced currents could affect the experiment, the low temperature thermometry and the sensitive microwave equipment.

3.2 Noise temperature of the amplification chain

The efficient measurement of very weak signals is of utmost importance for every experimental effort in CQED. Single photons in the microwave range must be amplified by as much as 90 dB to be able to detect them with commercial measurement devices. As discussed in Sec. 3.1, low temperature, low noise HEMT amplifiers are the first link of the amplifying chain. They add much more noise than the theoretical lower quantum bound for linear phase preserving amplifiers [Caves82] (for a recent review on quantum limited amplification, see [Clerk10]). HEMT's have, however, several practical advantages over quantum limited amplifiers, such as high gain ($\approx$ 30 dB), very large dynamic range (from single photons up to -30dBm input power), broadband operation of several GHz and a simple and stable operation. The basic properties of amplification chains are reviewed and the noise temperature of the implementation discussed in Sec. 3.1 is assessed.

3.2 Noise temperature of the amplification chain

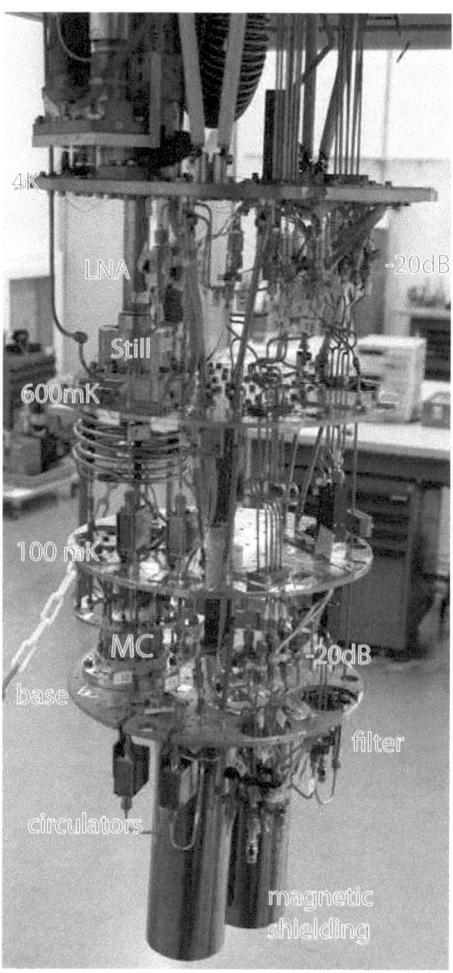

Figure 3.2 – Picture of the fully wired Vericold dilution refrigerator DR200. The input microwave cabling is visible on the right with its thermally anchored attenuators labeled '-20 dB'. The samples are shielded from magnetic fields by a double layer of high permeability metal (cryoperm). For the output paths, two lines with two circulators each before the two LNA are visible.

CHAPTER 3. EXPERIMENTAL SETUP

The amount of noise added to a signal by an amplifier can be expressed by several equivalent notations. The noise figure NF of an active device with gain G quantifies the ratio of the total noise power P_n^{out} delivered into the output load to the noise power P_n^{in} engendered by the input termination [IEE00]

$$NF = 10\log\left(\frac{P_n^{out}}{G \cdot P_n^{in}}\right). \tag{3.1}$$

It is defined at a given frequency of the device, when the noise temperature of its input termination is standard (290 K) and is expressed in dB. The noise factor F, is defined in an analog way

$$F := \frac{P_n^{out}}{G \cdot P_n^{in}} = 10^{\frac{NF}{10}}. \tag{3.2}$$

To understand the noise emitted by an amplifier, it is useful to separate the different noise sources in two categories: the noise delivered to the amplifier input and the noise added by the amplifier itself. The additional noise is assumed to be generated by an ideal load at the amplifier input, having a given noise temperature T_n, which not necessarily corresponds to the physical temperature of the amplifier. The thermal radiation generated by this ideal black body is then amplified by the device. So one can write the added power per Hertz of bandwidth at the input of an amplifier as [Kerr99]

$$P_n = k_B T_n \left[\left(\frac{\frac{h\nu}{k_B T_n}}{e^{\frac{h\nu}{k_B T_n}} - 1} + \frac{h\nu}{2}\right)\right]. \tag{3.3}$$

In the usual classical limit of $k_B T_n \gg h\nu$, Eq. (3.3) can be approximated by the well known Rayleigh-Jeans law

$$P_n = k_B T_n. \tag{3.4}$$

In the limit where Eq. (3.4) is valid, the noise temperature can also be expressed in terms of the noise factor

$$T_n = 290(F - 1). \tag{3.5}$$

If k amplifiers are added in series, each having a different gain G_i and noise factor F_i, the total noise factor F can easily be calculated and is given by

$$F = F_1 + \sum_{i=2}^{k} \frac{F_i - 1}{\prod_{j=1}^{i-1} G_j}. \tag{3.6}$$

3.2 Noise temperature of the amplification chain

Usually, $\prod_{j=1}^{i-1} G_j \gg (F_i - 1)$ so that the first amplifier defines the noise factor of the entire chain.

To characterize the noise properties of a system it is crucial to know the total gain as stated in Eq. (3.2). This is not simple in a cryogenic environment because the measured room temperature figures change when the components are cooled down. Usually, one uses a temperature controlled, calibrated noise source, as demonstrated in [Castellanos08], a fully isolated controller for a low temperature mechanical switch is discussed in Appendix A.6. Since a switchable resistor was not built in our setup, the qubit was employed to asses independently the absolute photon number induced in the resonator by a coherent tone. The AC-Stark shift experiment [Schuster05] indicated 1 photon in average in the resonator for a generated external tone with -35.1 dBm power for the wiring of the original wet Oxford cryostat, listed in Tab. 3.1. To ensure a stable operation of the amplifier, a low noise power supply was developed and implemented by the Elektronik-Lehrlabor (ELL) at ETH, see Appendix A.5. Using input-output theory [Gardiner85, Walls94] to relate the power in the resonator to the power transmitted through the cavity, results in

$$P_{signal} = n\hbar\omega_r \kappa, \quad (3.7)$$

where n is the number of photons in the resonator, $\omega_r/2\pi = 6.440$ GHz the resonance frequency and $\kappa/2\pi = 1.7$ MHz the photon decay rate. Calculating the power leaking out from the resonator with Eq. (3.7), leads to $P_{signal} = -141.4$ dBm. The calculated signal power, found adding up all components listed in Tab. 3.1 is $P_{signal} = -146$ dBm, where the difference of +4.6 dB is likely coming from the better performance of the stainless steel coaxial cables when cooled to low temperatures.

Before reaching the first cold amplifier, the signal transmitted through the resonator is attenuated by 7.9 dB (at room temperature). The amplified signal (38 dB gain expected) is then attenuated again by 10.2 dB before reaching the first warm amplifier. At this point the spectrum is accessible and was measured, finding -112.7 dBm for an applied -35 dBm resonant tone at the room temperature resonator input, while the noise was under the noise floor of the spectrum analyzer. The expected signal summing up all components is -125.2 dBm. 6 dB less attenuation each way for the cold cables/components are plausible and could again account for the difference.

After the two warm amplifiers, shown in Fig. 3.6, with 43 and 139 K noise temperatures

CHAPTER 3. EXPERIMENTAL SETUP

respectively and which amplify the signal by a total of 58.0 dB, the coherent tone has a power of -54.8 dBm. At this stage the noise of the system is measurable. One measures -106 dBm Hz noise power, showing the expected scaling with bandwidths. Using the nominal values for the components between sample holder and the first amplifier, implying a loss of around 8 dB, the system noise temperature is $T_n = 9$ K. With the big uncertainties in the cold attenuation factors discussed above, this is only a reference value. With 3 dB more or less attenuation, one would find $T_n = 6$ and 14 K respectively.

These figures can be improved deploying a cold amplifier with a smaller noise temperature or minimizing the losses between the sample holder and the first amplifier. The amplifier cannot be mounted closer the sample in the cryostat because of its power dissipation but the cables can be exchanged by superconducting lines which have similar electrical properties as the copper cables and thermal properties as the stainless steel ones.

A complete setup with two equal superconducting coaxial cables, leading to two separate cold amplifiers was implemented for the measurement of the correlation function of a single photon source in the microwave regime [Bozyigit10].

3.3 Amplitude and phase modulation of the control signal

The manipulation of the qubit in the time-domain requires the generation of fast RF pulses with accurate and precise control over frequency, phase and amplitude of the pulse. The direct generation of signals in the GHz frequency range with controlled phase and risetimes of the order of 1 ns cannot be accomplished with a single device. To circumvent this limitation the signal of an arbitrary waveform generator with 1 ns resolution is upconverted to the frequency generated by a low noise analog microwave generator ω_{LO} using an IQ-mixer, as sketched in Fig. 3.1. If the upconverting signal is modulated, any frequency around ω_{LO} (in the bandwidth of the mixer) can be generated. Using an IQ-mixer instead of a simple mixer allows to control the phase of the signal.

An ideal IQ-mixer can be considered as composed by two balanced mixers and two hybrids which multiply the signal applied to the local oscillator port s_{LO} with the in-phase signal s_I and add it to the multiplication of the quadrature signal s_Q with s_{LO}, phase shifted by $\pi/2$, as shown in Fig. 3.3. The high frequency local oscillator is driving the mixer electronics and must

3.3 Amplitude and phase modulation of the control signal

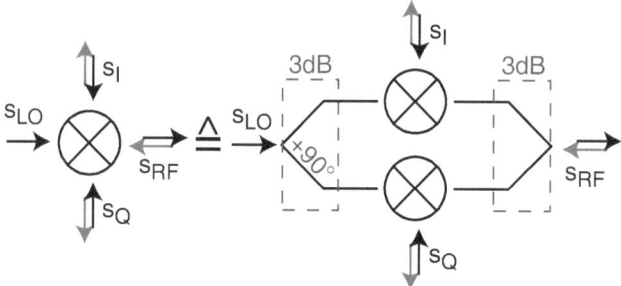

Figure 3.3 – Schematics of an IQ-mixer modeled as a 90° hybrid followed by two mixers and a final symmetric hybrid to rejoin the signals. The signal paths leading to upconversion are indicated in black, while the downconversion is shown in red.

therefore have a fixed amplitude A_{LO}. We use this tone as a reference, defining its phase zero and frequency ω_{LO}. The I and Q quadrature signals are assumed to have a fixed frequency $\omega_{I/Q}$ but variable phases $\phi_{I/Q}(t)$ and time dependent amplitudes $A_{I/Q}(t)$. s_Q has an additional constant phase φ, which will be useful to select the generated sideband. This definitions yield

$$\begin{aligned}
s_{LO} &:= A_{LO}\cos(\omega_{LO}t), \\
s_I &:= A_I(t)\cos(\omega_I t + \phi_I(t)), \\
s_Q &:= A_Q(t)\cos(\omega_Q t + \phi_Q(t) - \varphi), \\
s_{RF} &= \frac{A_{LO}}{\sqrt{2}}\left[\cos(\omega_{LO}t)\cdot s_I + \cos(\omega_{LO}t + \pi/2)\cdot s_Q\right].
\end{aligned} \qquad (3.8)$$

Using simple trigonometric relations and assuming $A_I(t) = A_Q(t) \equiv A_{IQ}(t)$, $\omega_I = \omega_Q \equiv \omega_{IQ}$ and $\phi_I(t) = \phi_Q(t) \equiv \phi_{IQ}(t)$, it follows that

$$\begin{aligned}
\varphi = 0 \;\rightarrow\; s_{RF} &= \frac{A_{LO}A_{IQ}(t)}{2}\{\cos\left[(\omega_{LO}+\omega_{IQ})t + \phi_{IQ}(t) - 3\pi/4\right] \\
&\quad + \cos\left[(\omega_{LO}-\omega_{IQ})t - \phi_{IQ}(t) - 3\pi/4\right]\}, \\
\varphi = \pi/2 \;\rightarrow\; s_{RF} &= \frac{A_{LO}A_{IQ}(t)}{\sqrt{2}}\cos\left[(\omega_{LO}+\omega_{IQ})t + \phi_{IQ}(t)\right], \\
\varphi = -\pi/2 \;\rightarrow\; s_{RF} &= \frac{A_{LO}A_{IQ}(t)}{\sqrt{2}}\cos\left[(\omega_{LO}-\omega_{IQ})t - \phi_{IQ}(t)\right].
\end{aligned} \qquad (3.9)$$

CHAPTER 3. EXPERIMENTAL SETUP

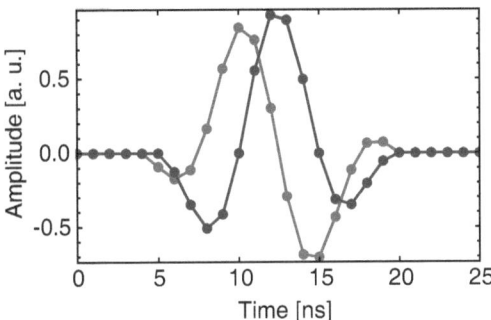

Figure 3.4 – s_I (red) and s_Q (blue) used to generate a gaussian pulse at the frequency $\omega_{LO}+\omega_{IQ}$. This particular implementation has $\sigma = 4$ ns and $\omega_{IQ}/2\pi = 100$ MHz. Note that $s_{I/Q}$ are 90 degrees off phase to generate a pulse on a single sideband, as stated in Eq. (3.9). The used arbitrary waveform generator (Tektronix AWG5014) has a time resolution of 1 ns, reflected in the figure by the supporting points.

Adjusting the free parameter φ, a single mixer can generate identical signals on both sidebands $\omega_{LO} \pm \omega_{IQ}$ of the local oscillator, like a normal mixer or selectively suppress one sideband generating a single tone at $\omega_{LO} + \omega_{IQ}$ or $\omega_{LO} - \omega_{IQ}$. The amplitude of the selected sideband can be modulated with $A_{IQ}(t)$, and the phase changed using $\phi_{IQ}(t)$, which is alike a frequency modulation. This flexibility is useful if one needs to generate pulses on different transitions of the same qubit using a single generator and mixer. Moreover, the implementation of a modulated IQ signal avoids the upconversion of 1/f noise and unwanted resonant leakage. An exemplar gaussian pulse is shown in Fig. 3.4.

These IQ-mixers usually do not behave ideally. They are affected by considerable DC-offsets, leading to substantial leakage of s_{LO} into s_{RF} even if $s_{LO} = s_{LO} = 0$ and phase and amplitude imbalance which generate substantial signals at unwanted frequencies (only about 20 dB lower than the wanted tone without further calibration). It is, however, possible to calibrate out these nonlinearities and obtain an isolation of more than 40 dB. The calibration has to be repeated for different ω_{LO} frequencies as well as for different modulation frequencies ω_{IQ}. The magnitude of the generated signal A_{RF}, needed for example to compare Rabi oscillation rates at different detunings, as performed in Sec. 5.5, depends on the selected ω_{LO} and ω_{IQ} and must therefore be measured each time. To avoid the presence of higher order sidebands typically at $\omega_{LO} \pm 2\omega_{IQ}$, it is necessary to drive the mixer with low enough $s_{I/Q}$ amplitudes (typically

3.3 Amplitude and phase modulation of the control signal

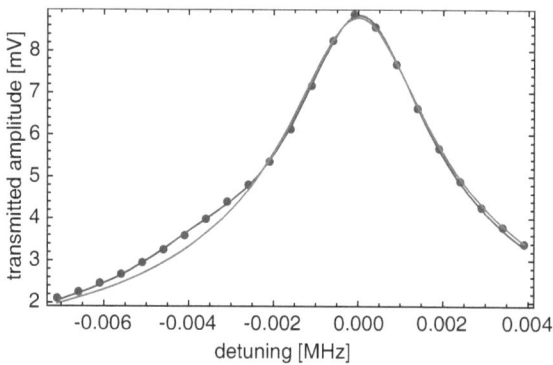

Figure 3.5 – Resonator spectrum weakly probed with a drive corresponding to about 1 photon in average in the cavity on resonance (blue dots). The red line is the best Lorentzian fit to the data, while the blue curve considers a second Lorentzian peak, finding a detuning of 3.6 MHz, corresponding to $2\chi/2\pi = 3.2$ MHz. The side peak amplitude of $5 \pm 1\%$ of the main peak enables for a rough estimation of the incoherent population of the excited state.

400 mV), to avoid its nonlinear response.

The additional installation of attenuation between the AWG output and the mixer IQ ports is helping inhibiting broadband noise being upconverted to the resonant qubit transition inducing unwanted populations of the order of 5 %. This population is evaluated measuring the resonator spectrum without any explicit qubit drive, as shown in Fig. 3.5. The non-Lorentzian response is due to a small incoherent excited state population of the qubit which shifts the resonator by 2χ, see Sec. 2.4. The average over many realizations leads to two separate Lorentzian peaks with amplitudes weighted by the populations, as observed in Fig. 3.5. Switching off the microwave tone s_{LO}, using the built in pulse function to avoid such problems is not practical because the generator is too slow. The risetime of about 10 ns, combined with fluctuations lasting up to 100 ns affect the quality of the pulses in such a way that this technique is not used.

CHAPTER 3. EXPERIMENTAL SETUP

3.4 Measurement signal demodulation

Information on the qubit state and on the field inside the cavity can be gained by measuring the quadrature amplitudes of the electromagnetic wave leaking out of the resonator through the output capacitance C_{out}, as described in Sec. 2.4. Note that only half of the total signal amplitude is transmitted through C_{out} (in the case $C_{out} = C_{in}$), while the other half is lost via C_{in}, which could be accessed using an additional circulator placed just before C_{in} and a second measurement line. A design with $C_{out} \gg C_{in}$, would ensure almost every photon leaving the cavity from C_{out}, doubling the signal (considering a constant κ). The implementation of a second measurement line, monitoring the reflected signal in the case $C_{out} = C_{in}$ would, however, not lead to an increased signal to noise ratio, since a second amplifier would double the noise.

Operating the cavity in the few photons regime requires many repetitions of the same experiment to average out the dominating amplifier noise discussed in Sec. 3.2. A direct measurement of the transmitted signal could be performed with an high frequency oscilloscope, which can, however, not average efficiently. An analog-to-digital converter (ADC) combined with a field-programmable gate array (FPGA) on a computer card is much more versatile. The microwave signal must be downconverted to be within the limited bandwidth of the card, limiting the bandwidth of the measurement to the downconverted frequency. The acquired signal can, however, be manipulated in real time and diverse operations such as averaging, fourier transformation and arbitrary correlation measurements can be performed efficiently on a single device. Commercial FPGA based acquisition cards usually also feature digital-to-analog converters (DAC) which could be used to implement real time feedback, enabling error correction algorithms [Knill00], quantum teleportation [Bennett93] without the need for post-selection or measurement based state preparation [Bishop09b].

A time-resolved, phase sensitive measurement of the transmission quadrature amplitudes is realized by down converting the measurement signal s_{RF} using an IQ-mixer in an inverted way compared to Eq. (3.8), as indicated by the red arrows in Fig. 3.3. s_{RF} is multiplied with s_{LO}, generating two phase shifted outputs $s_{I/Q}$, which are both needed to reconstruct simultaneously the phase and amplitude (or I and Q quadratures) of the original signal if one downconverts to DC ($\omega_{RF} = \omega_{LO}$). The hardware used is sketched in Fig. 3.1 and shown in Fig. 3.6. The downconversion to an intermediate frequency $\Delta_{RF,LO} = \omega_{RF} - \omega_{LO} = 25$ MHz and the successive digital downconversion to DC avoids the requirement to simultaneously measure both channels of the IQ mixer. What is more important, DC-offsets do not carry any

3.4 Measurement signal demodulation

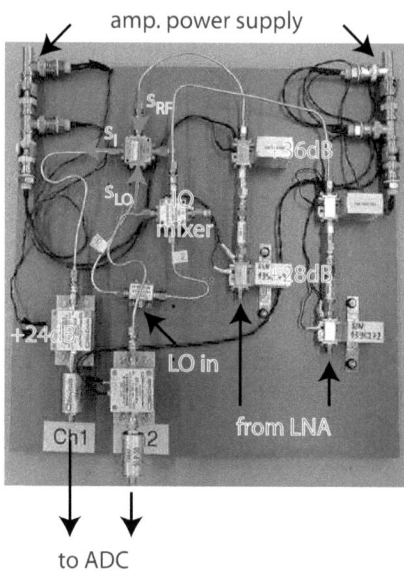

Figure 3.6 – Downconversion board for the measurement signal. The signal from the cold amplifier is first further amplified, then filtered (to avoid a saturation of the next amplifier by the broadband noise), again amplified, downconverted to 25 MHz using an IQ mixer and finally amplified and filtered with a low pass filter before being digitized.

information and can therefore be neglected, avoiding 1/f noise. The limited bandwidth of this method of 25 MHz still exceeds the typical resonator bandwidth, yielding one independent data point every 40 ns. A detailed description of the digital downconversion and filtering procedure can be found in [Schuster07a, Lang09].

A typical time-resolved experiment is repeated every 10 μs and averaged 65'000 times to enhance the signal-to-noise-ratio. It is then digitally down converted to DC, leading to a total measurement time of 650 milliseconds for 250 points in each of the traces $s_{I/Q}(t)$. Using input-output theory [Gardiner85, Walls94], the measured, averaged signal quadratures $s_{I/Q}(t)$ at the

CHAPTER 3. EXPERIMENTAL SETUP

output of the resonator are related to the field inside the cavity by

$$s_I(t) = \sqrt{Z_0 \hbar \omega_r \kappa} \, \Re\langle \hat{a}(t) \rangle,$$
$$s_Q(t) = \sqrt{Z_0 \hbar \omega_r \kappa} \, \Im\langle \hat{a}(t) \rangle, \qquad (3.10)$$

where Z_0 is the characteristic impedance of the transmission line connected to the resonator. The reflected amplitudes can be calculated similarly, taking into account both the signal leaking out from the cavity and the signal reflected at the resonator input capacitance. From this relations one can reconstruct the field inside the cavity which in turn carries information about the qubits coupled to the resonator, enabling their state reconstruction, as discussed in chapter 4.

3.5 Selective DC flux control

As discussed in Sec. 2.2, the qubit transition frequency can be tuned via an externally applied flux Φ, threading the superconducting loop. The typical on chip separation of the qubits is of the order of several millimeters. For this reason it was possible to design and manufacture a set of small coils, mounted directly on the sample-holder able to tune three different qubits independently. The fast on chip flux lines, seen in Sec. 3.3, have limited tunability due to restrictions on the maximal current flowing without affecting the sample temperature or stability. Moreover a DC current can be filtered efficiently avoiding unwanted $1/f$ noise that affects the phase coherence of the qubits.

For a typical qubit SQUID-loop size of 2 μm^2, a perpendicular field of around 1 mT is necessary to tune the qubit by 1 period, corresponding to 1 flux quantum $\Phi_0 = h/(2e)$. To reach such a field with coils small enough to address separate qubits, they have to be close to the sample. Figure 3.7 shows two small coils with an external radius of 3.5 mm, milled in the backplate of the sample-holder and a third, bigger coil added on top. 67 m and 357 m of a 36 μm thick superconducting wire (from Supercon Inc., Boston USA, wire type SC-T48B-M-0.025mm) is used to wind 3600 and 7000 turns on the small and big coils, respectively. The superconducting wire is needed to avoid an excessive heat load on the baseplate. For the designed current of 1 mA, even an unrealistic small resistance of 1 Ohm, for normal metallic wires with this length, would already significantly heat up the cryostat. The inductance was measured with a simple voltage meter using the included inductance function, finding 0.05 and 0.77 H respectively, in good agreement with the expected values from the Weeler formula.

3.5 Selective DC flux control

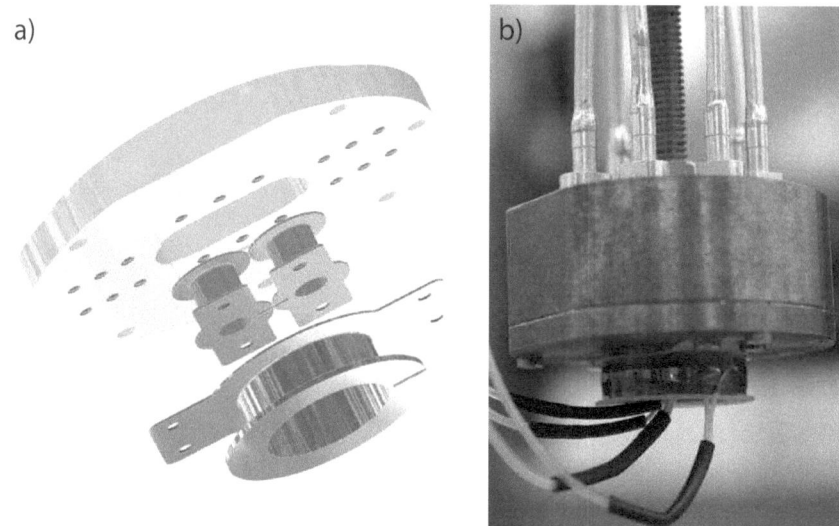

Figure 3.7 – a) Cad sketch of the coil-holders shown below the back of the sample-holder lid. These are used to independently tune the flux of three separate qubits. b) Image of a loaded sample-holder with mounted coils and feed lines.

First generation coil-bodies were made from copper, to ensure an adequate thermalization of the coils. The thin wire isolation was, however, prone to damages, resulting in shorts to the body. To avoid this problem, newer coil-bodies are made of Stycast®1266 epoxy which also thermalizes at millikelvin temperatures [Armstrong78] and has proven to sustain thermal cycling.

The calculated field at the sample location (1 and 6 mm from the coil-bodies ends) for a current of 1 mA is 0.37 and 0.20 mT for the small and big coil respectively and is in good agreement with a measurement performed with a small hall sensor, performed at room temperature. The small coils generate a strong field only for qubits located immediately below their axis, with a considerably smaller field for qubits placed few millimeters away. The big coil, however, adds a constant and global field over the entire chip. Flux focussing due to the superconducting ground planes increases the field at the qubit location by a factor up to 50. The effective coupling strengths can therefore be substantially different in dependence of the

sample geometry.
If different qubits show different coupling strengths to different coils, any combination of fluxes can be chosen. This is desirable to tune the different qubit transition frequencies to the needed values. Since the field of the coils are additive, it is possible to solve the linear equation

$$\begin{pmatrix} \phi_a \\ \phi_b \\ \phi_c \end{pmatrix} - \begin{pmatrix} \phi_a^{off} \\ \phi_b^{off} \\ \phi_c^{off} \end{pmatrix} = \begin{pmatrix} f_{00} & f_{01} & f_{02} \\ f_{10} & f_{11} & f_{12} \\ f_{20} & f_{21} & f_{22} \end{pmatrix} \begin{pmatrix} I_{small}^1 \\ I_{big} \\ I_{small}^2 \end{pmatrix}, \quad (3.11)$$

to find the currents I_i needed for the chosen fluxes ϕ_i. Formally, the coupling matrix f_{ij} has to be nonsingular such that a solution of Eq. (3.11) exists for arbitrary fluxes ϕ_i.

As discussed in Sec. 2.2, the qubit transition frequency depends on the flux ϕ threading the SQUID-loop as

$$\omega_0 \propto E_{J,max} \cos\left(\frac{\pi\phi}{\phi_0}\right). \quad (3.12)$$

Assuming a linear relation between the current in a specific coil and the flux in a given qubit, it is therefore possible to infer the current needed for a given flux. This is done spectroscopically, where the qubit transition frequency is measured in function of the applied current and the periodicity is fitted, as shown in Fig. 2.6 and Fig. 2.7. The coupling strengths f_{ij} are then the inverse of the measured periodicity of the qubit transition frequencies in the applied current and $\phi_i^{off} \approx 1 - 10$ mΦ_0 are small offsets due to an imperfect magnetic shielding of the sample. The experiment depicted above is repeated with every possible qubit/coil combination to determine all f_{ij} elements. Typically, the periodicity of the big coil is of the order of 0.2 mA for all qubits, while the small coils show a stronger coupling, with a periodicity of around 0.1 mA for the qubits on axis and a cross coupling to the off axis qubits of -1/4.7 mA. The high degree of control achieved using these coils is demonstrated in [Fink09].

Low frequency current noise in the coils couples directly to the qubit transition frequency causing dephasing [Abragam61, Ithier05]. For this reason, careful low-noise biasing has to be implemented avoiding in particular 50 Hz noise from the power grid and 1/f noise. The cryogenic DC wiring also has to minimize sources of noise, such as inductive coupling. The currents induced by a current loop moving in a magnetic field are minimized using twisted pair wires, so that current pickup in combination with vibrations e.g. from the pulse tube cooler are suppressed.

To eliminate the ubiquitous 50 Hz noise, a simple first order RC filter, sketched in Fig. 3.8a

3.5 Selective DC flux control

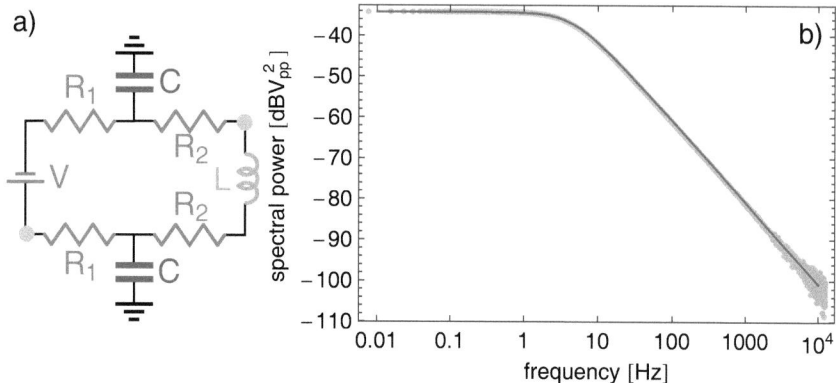

Figure 3.8 – a) Circuit of the low pass filter used to voltage bias the superconducting coils. b) Measured spectral power density through one branch of the filter shown in a) ($C = 6.8$ μF, $R_1 = 5$ kOhm, $R_2 = 3$ kOhm $\ll$ $R_{\text{FFTanal.}} = 10$ MOhm).

has been implemented. It simultaneously acts as a bias resistor, enabling to generate the needed current in the superconducting coil (L) with a programmable voltage source. The two R_1 and C act as a filter on both leads while R_2 ensures a high impedance environment on the filter output. The behavior of a single branch of the filter is shown in Fig. 3.8b, measured with a low frequency, high impedance FFT analyzer between the two cyan dots. A symmetric design was implemented to avoid any noise propagating to the coil from one side and a single branch was measured to ensure the working of the filter independently from the other branch. The calculated filter response is shown as a solid line in the same plot. The total resistance $2R_1 + 2R_2$ is chosen to exhaust the full range of the voltage source, while the ratio R_1/R_2 is adjusted to optimize the filter cutoff, ensuring good suppression of 50 Hz noise while maintaining a reasonable RC time constant τ of around 0.1 s. The DC offset in the noise power spectrum shown in Fig. 3.8b is due to the resistors which act as voltage dividers.

The relative flux variation induced by a 1 V fluctuation at the input of two different filters is shown in Fig. 3.9a, calculated with the full network shown in Fig. 3.8a using Kirchoff Kirchhoff's circuit laws

$$V_L(\omega) = \frac{R_2 V}{(1 + iC_1 R_1 \omega)\left[-2(R_1 + R_2) - i\omega(L + 2R_1 R_2 C_1) + C_1 R_1 L \omega^2\right]}. \tag{3.13}$$

CHAPTER 3. EXPERIMENTAL SETUP

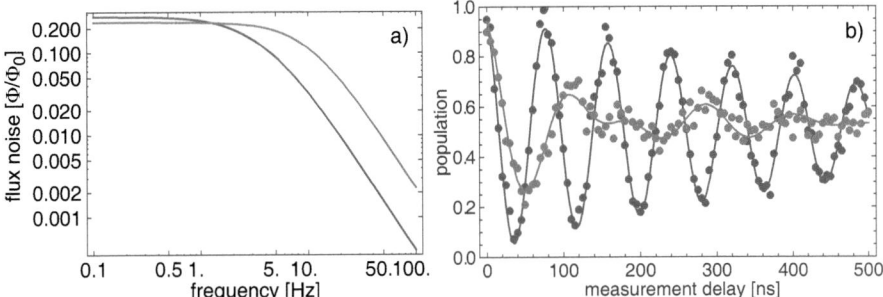

Figure 3.9 – a) Calculated relative flux noise amplitude induced by a 1 V fluctuation for the filter used for the big coil ($C = 6.8$ µF, $R_1 = 5$ kOhm, $R_2 = 3$ kOhm in red) and small coils ($C = 6.8$ µF, $R_1 = 11$ kOhm, $R_2 = 5$ kOhm in blue). b) Measured Ramsey oscillations, driven with 10 MHz detuning, using only bias resistors on a Yokogawa DC 7651 source (red) and the discussed low pass filter (blue). The blue points are fitted to an exponentially decaying oscillation, while the red dots are fitted to two different frequencies.

V_L is the voltage drop across the coil induced by the voltage V applied to the filter and can be related to a flux if the flux periodicity of the considered coil is known. A measured flux periodicity of 3.6 and 4.2 V/ϕ_0, respectively is used to calculate the flux noise in Fig. 3.8a. The 3 dB point is at 2 and 6 Hz for the small and big coil respectively, implying a waiting time after each voltage change of 1 s to reach 99% of the new current. The flux noise at 50 Hz is suppressed by a factor of 150 and 30 respectively (in addition of the voltage division). The big inductance of the coils does not improve the filter behavior below 1'000 Hz.

To asses the performance of the filters, Ramsey oscillations, see Sec. 4.7, were measured to analyze the phase coherence of the qubit. The data taken with the filters is shown in Fig. 3.9b in blue and displays the usual exponential decay with $T_2 = 500$ ns. The same result is obtained without any filter but using a battery as voltage source. Using simple symmetric bias resistors with a Yokogawa DC 7651 source results in the red dots. The coherence is clearly reduced and the oscillations show beating, implying the presence of a second frequency. This cannot be explained with broadband noise and must be ascribed to a coherent modulation of the qubit transition frequency, most likely at 50 Hz. The data is fitted to two tones with 10 and 20 MHz respectively and a decay time of 150 ns. A first generation of filters, with a single central capacitance displayed the same issue.

3.6 Sample fabrication

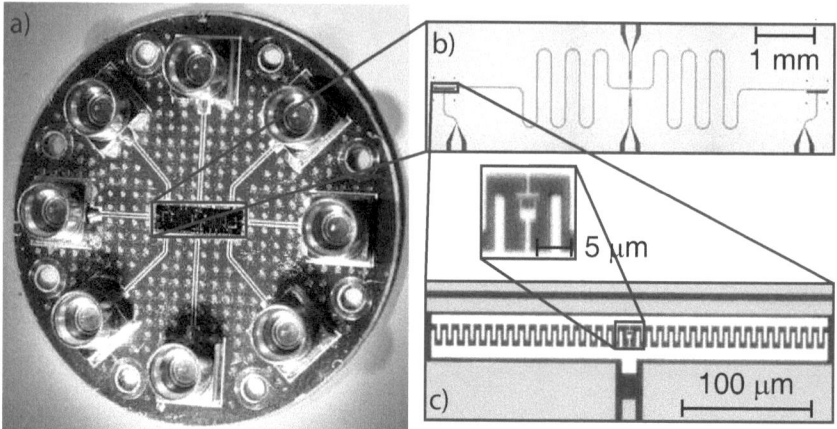

Figure 3.10 – a) Circuit QED device mounted onto a high frequency sample-holder featuring 8 high frequency coplanar waveguides. The printed board circuit has a diameter of 30 mm. b) False color optical microscope image of a multi-Q resonator sample [Leek10]. c) Zoom-in of a transmon type qubit with further enlargement of the superconducting loop.

The filters are able to suppress the pronounced 50 Hz noise from the Yokogawa source, but to avoid any future problems, it was decided to use only battery powered voltage sources (SRS SIM928), in addition to the filters.

3.6 Sample fabrication

The resonators are fabricated using conventional optical lithography, while the qubits are written with electron beam lithography and grown using a shadow evaporation technique. All samples measured in this thesis have been fabricated in the cleanroom facility for advanced micro and nanotechnology at ETH (FIRST). The processes were mainly developed by Martin Göppl and are described in detail in his Phd thesis [Göppl09]. Here, only the main techniques used for the fabrication are briefly discussed.

For the fabrication of high quality resonators, a 150 nm thick layer of niobium is sputtered on a sapphire substrate. A film of photoresist is spun and baked on top of it and then exposed to UV light through a patterned mask carrying the resonator design. The exposed resist is then developed and the uncovered metal layer is removed by reactive ion etching. A typical false color microscope photograph of a chip, 7 × 2 mm in size, is shown in Fig. 3.10b. Niobium film is shown as white and the sapphire substrate in green. The resonator is meandered to reach the desired length within the limited chip space. At the left and right end of the chip there is space to accommodate a transmon type qubit with a charge gate line for the selective qubit control.

The qubits are fabricated in a separate step. Two thin films of electron beam resist are spun and baked on a chip which already hosts a resonator, before being directly written with an electron beam with an appropriate pattern. After development, a free standing bridge of photoresist is generated due to the higher sensitivity of the lower layer of resist which is also exposed more due to electrons backscattered from the sapphire substrate. A thin layer of aluminum is evaporated at a specific angle, forming the bottom electrode of the Josephson junction. Afterwards, the sample is exposed for a controlled length of time to an oxygen containing gas mixture which controllably oxidizes the aluminum. This forms the tunnel barriers for the junctions. In a subsequent step, a second layer of aluminum is evaporated at a different angle, generating a partial overlapping region which defines the Josephson junction. In a final step the photoresist and the remaining metallization layer are lifted off. A resulting transmon type qubit is shown in Fig. 3.10c with its two big interdigitated metallic plates forming the shunt capacitor C_s. The superconducting loop (~ 2 μm^2) with its two small junctions ($\sim 100 \times 100$ nm) is shown in the center of Fig. 3.10.

Finally the chip is mounted on a printed board circuit (PCB) which is connected to the cryogenic cables via commercial SMP connectors, see Fig. 3.10a. The different conductors are wire-bonded to the PCB and unwanted microwave frequency resonance modes are suppressed using wire bonds across the resonator.

3.7 Power dependence of high Q resonators

To achieve long coherence times, needed for any practical implementation of a quantum processor or memory, the sources of decoherence have to be carefully assessed and minimized. The photon loss to the feed lines, through the capacitors (Sec. 2.1) can be changed during design [Göppl08], while a generic and unknown internal loss mechanism is intrinsically given. Any source of decoherence within the cavity, however, couples strongly to the artificial atoms and causes unwanted noise. For this reason a quantitative understanding of the losses and their

3.7 Power dependence of high Q resonators

sources in the few photon regime is desirable [Gao06, Gao07, O'Connell08, Kumar08, Gao08b, Chen08, Gao08a, Barends08, Wang09b, Barends10]. High quality resonators are therefore studied at low temperatures and low excitation powers and the loss mechanisms are assessed.

In a linear system the photon number n in the resonator can be calculated with a simple energy conservation argument. In the steady state and on resonance, an incoming traveling voltage wave with amplitude V_{in} generates a standing wave in the resonator. The incoming voltage wave is completely reflected and phase shifted by π at the resonator input capacitance ($V_{ref} = -V_{in}$). For a symmetrically coupled resonator, the voltage leaking out from the cavity is identical on both sides ($V_{out,L} = V_{out,R} = V_{out}$), so the energy conservation law implies

$$\begin{align} P_{in} &= P_{ref} + P_{trans} + P_{loss} \tag{3.14a} \\ &= (V_{ref} + V_{out,L})^2 + V_{out,R}^2 + P_{loss} \tag{3.14b} \\ &= (\sqrt{P_{in}} - \sqrt{P_{out}})^2 + P_{out} + P_{loss}. \tag{3.14c} \end{align}$$

Introducing the coupling coefficient $g := Q_{int}/Q_{ext}$, and inserting Eq. (2.4) into Eq. (3.14), the insertion loss (IL) can be expressed as follows

$$\begin{align} IL := \frac{P_{out}}{P_{in}} &= |S_{21}|^2 = 1 \Big/ \left(1 + \frac{P_{loss}^2}{4P_{out}^2} + \frac{P_{loss}}{P_{out}} \right) \tag{3.15a} \\ &= \left(\frac{g}{g+1} \right)^2, \tag{3.15b} \end{align}$$

where S_{21} is the scattering parameter voltage gain. The expression for the dissipated power can be evaluated with Eq. (3.15)

$$\frac{P_{loss}}{P_{in}} = 2\sqrt{\frac{P_{out}}{P_{in}}} - 2\frac{P_{out}}{P_{in}} = \frac{2g}{(g+1)^2}. \tag{3.16}$$

The reflected power is then found subtracting Eq. (3.15) and Eq. (3.16) from Eq. (3.14)

$$\frac{P_{ref}}{P_{in}} = |S_{11}|^2 = \frac{1}{(g+1)^2}. \tag{3.17}$$

The same results are obtained by a full analysis of the equivalent electric circuit [McKinstry89] or using the ABCD matrix formalism [Leong02]. The spectrum of the system was calculated numerically using an exact ABCD matrix approach finding no deviations from the results of

CHAPTER 3. EXPERIMENTAL SETUP

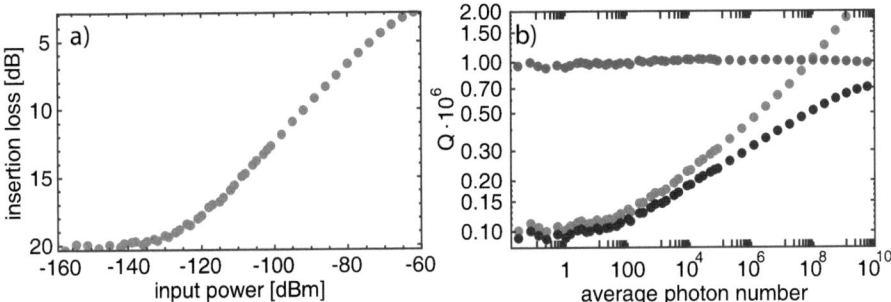

Figure 3.11 – a) Insertion loss fitted to the measured Lorentzian resonator responses versus measurement power. b) Measured loaded quality factor Q_L (black) versus photon number n reconstructed using Eq. (3.18). The internal Q_{int} (red) and external Q_{ext} (blue) quality factors are calculated with Eq. (3.15).

Eq. (3.15) and Eq. (3.17). Inserting Eq. (3.16) in Eq. (2.4) leads to the average number of photons n inside the cavity on resonance

$$n = P_{in}\left[\frac{2g}{(1+g)^2}\right]\frac{Q_{int}}{\hbar\omega^2} = 2P_{in}\sqrt{IL}\frac{Q_L}{\hbar\omega^2}. \quad (3.18)$$

From the measured insertion loss, loaded quality factor and applied power it is therefore possible to infer the average photon number in the resonator and the internal quality factor. Figure 3.11a shows the measured insertion loss, while b) displays the fitted Q_L of the high quality resonator cooled to about 20 mK versus the applied measurement power at the input of the resonator. The resonator is made of niobium on a sapphire substrate, with the fundamental resonance frequency at $\omega_r/2\pi = 7.135$ GHz and symmetric 10 μm gaps, implying a theoretical $Q_{ext} = 1.35\ 10^6$ (sample ID: Sa05.5.Z3). At high powers, the resonator is operated in the overcoupled regime, where $Q_{int} \gg Q_{ext} \approx Q_L$, indicated by the low insertion loss. For lower powers, Q_L decreases steadily until it saturates at low photon numbers. The constant Q_{ext}, observed in Fig. 3.11b, arises naturally from Eq. (3.15) and confirms the validity of the discussed method. The cold calibration of the insertion loss of the cabling leading to the resonator is accurate only up to about 2 dB, implying an uncertainty ΔQ_{ext} of 200'000.

This behavior has been observed in several experiments studying low power responses of high quality resonators [Gao06, Gao07, O'Connell08, Kumar08, Gao08b, Chen08, Gao08a,

3.7 Power dependence of high Q resonators

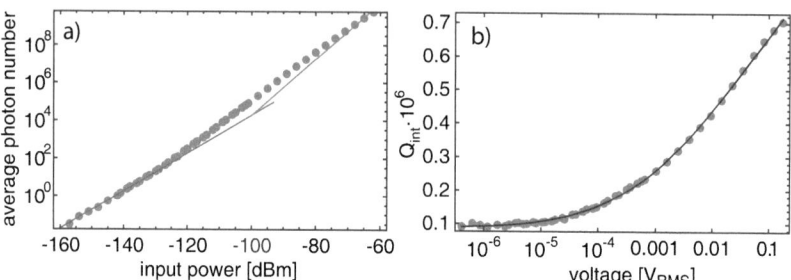

Figure 3.12 – a) Average photon number on resonance in the cavity versus driving power. The blue line indicates a linear dependence, typical for a system with constant quality factor b) Scaling of the internal quality factor versus the RMS voltage in the cavity. The black line is fitted to Eq. (3.19).

Barends08, Wang09b, Barends10] for different designs and material choices. The typical photonic temperature at GHz frequencies in the resonator is well below 100 mK [Fink10], which implies less than 0.04 photons in average in the cavity due to thermal fluctuations. The observed onset of the saturation of the quality factor at around 100 photons is therefore not due to thermally activated fluctuations.

The losses are usually explained by two level systems (TLS) which couple to the field and are saturated by high powers. Their exact origin and distribution is being actively investigated but recent results [Wang09b, Barends10] suggest their location on the metallic surface and not in the bulk of the dielectric, implying a scaling

$$Q_{int} = Q_{TLS}\sqrt{1 + (V_{rms}/V_{sat})^\beta}, \qquad (3.19)$$

where $V_{rms} = V_{in}[2gQ_L/(1+g)]^{1/2}$ is the root-mean squared voltage on the center conductor and V_{sat} the saturation field for the TLS and β a geometry dependent exponent. Figure 3.12a clearly shows the saturation of Q_{int} at low powers, where similarly to [O'Connell08], there is a linear relation between input power and photon number (blue line), while for higher powers the TLS are gradually saturated, resulting in a slope of around 3/2 (magenta line). The data shown in Fig. 3.12b is fitted to Eq. (3.18), finding $Q_{TLS} = 95'000$, $V_{sat} = 3 \; 10^{-5}$ V and $\beta = 0.7$, similar to the result found for example by [Gao07].

The TLS's which are probably limiting the photon lifetimes at low powers could also be

48

responsible of the limited lifetime of superconducting qubits. Further research in the fabrication of high quality resonators could therefore also help understanding the fundamental mechanisms of qubit relaxation.

Chapter 4

Dispersive Quantum State Readout

An indispensable prerequisite for each physical implementation of a quantum information processing system is the ability to read-out the quantum state of a specific qubit after each manipulation with high quantum efficiency [DiVincenzo97, DiVincenzo00]. For a given single qubit state $|\Psi\rangle = c_g|g\rangle + c_e|e\rangle$, an ideal strong measurement returns ground state "g" with probability $|c_g|^2$ and excited state "e" with probability $|c_e|^2$, independent of the state of other qubits or of the state of the environment. A measurement is said to be quantum non-demolition (QND) if after the measurement the qubit state is left in $|g\rangle$ after the "g" outcome and $|e\rangle$ after the "e" outcome [Walls94]. This property is not necessary for quantum computation since after the measurement the system can be reinitialized, but is a valuable feature if one has only a weak or noisy measurement because it allows for a subsequent state determination which enhances the efficiency. A measurement is said to be weak [Aharonov88] if the measurement apparatus couples only weakly to the qubit and therefore does not induce a full wave function collapse. The probability of getting the result 'the system is in the "g/e" state' is then

$$p_{g/e} = \frac{1}{2} + \delta \left(|c_{g/e}|^2 - \frac{1}{2} \right), \tag{4.1}$$

with an arbitrary small δ. In the limit $\delta \to 1$, we obtain the strong measurement. For small δ, however, the measurement efficiency m_e is lowered to [Nielsen00]

$$m_e := \frac{\frac{1}{2} + \delta \left(|c_{g/e}|^2 - \frac{1}{2} \right)}{|c_{g/e}|^2}. \tag{4.2}$$

Such a weak measurement or a state determination with low efficiency due to instrumental noise for example, is still sufficient for arbitrary quantum computation. If a specific task requires a higher efficiency m_e, the measurement is repeated or averaged over an ensemble of qubits to reach higher values.

For superconducting qubits, a number of read-out strategies specific to various implementations have been pursued. The first demonstration of coherent superposition of quantum states in a macroscopic solid state superconducting system was performed by Nakamura and coworkers in 1999 [Nakamura99]. They considered a charge-qubit realized with a Cooper-pair box structure and read-out the state measuring the current through a probe junction directly connected to the qubit, biased to measure single quasi particle tunneling. For charge qubits the charging energy E_C dominates over the Josephson energy E_J, see Sec. 2.2. Different states are characterized by a well defined number of Cooper-pairs localized on the island. A single electron transistor (SET) [Devoret00] is a perfect device to detect this charge and therefore distinguish the states [Duty04, Astafiev04]. The strong current noise back-action from the SET will, however, limit the qubit lifetime and interfere with other nearby qubits. A related type of charge-qubit, the quantronium, has a third big Josephson junction added to the qubit loop which is used as a read-out device [Vion02]. In this configuration the charging energy E_C and the Josephson energy E_J have similar magnitude and the number of charges on the island is no more a good quantum number. To read-out the state, a current is applied to the big junction and the probability of switching to the finite voltage state is recorded.

The read-out of early flux-qubits consisting of a loop interrupted by three Josephson-junctions has been performed by a nearby direct current superconducting quantum interference device (DC-SQUID) [vanderWal00, Chiorescu03, Chiorescu04]. For these devices E_J is much bigger than E_C and each state is characterized by a well defined current circulating in the loop. The different values of the flux in the DC-SQUID produced by the persistent currents of states are discerned by different bias currents at which the SQUID switches from the supercurrent branch to a finite voltage state. The switching process, however, produces quasiparticles, which in turn can interact with nearby qubits in an uncontrolled way.

A third type of qubit, the phase qubit, is based on current biased Josephson junctions and has similar E_J to E_C ratio as a flux qubit, but is read-out in a different way. The different states have different probabilities of tunneling out from the potential and of building up a voltage across the junction which is in turn measured [Martinis02, Simmonds04]. The tunneling of the state implies a destructive measurement and produces dissipation, unwanted in a system with

CHAPTER 4. DISPERSIVE QUANTUM STATE READOUT

many qubits.

To avoid these problems, different groups coupled the qubit to an harmonic oscillator acting as a non dissipative read-out device. It was demonstrated for a cooper pair box coupled to a lumped element circuit [Sillanpää05], for a flux qubit coupled to a tank circuit [Grajcar04] and for a flux qubit dispersively coupled to a resonator [Lupascu04, Lupascu05]. To improve the measurement efficiency the intrinsic nonlinearity of the SQUID resonator was used to perform a latching measurement [Lupascu06] and demonstrated the QND nature of the readout [Lupascu07]. In a similar approach [Siddiqi06], a quantronium type qubit was combined with a Josephson bifurcation amplifier [Siddiqi04].

In the circuit quantum electrodynamics architecture, a qubit is strongly coupled to a high quality transmission line resonator [Blais04, Wallraff04]. The read-out is accomplished by detecting the dispersive qubit state-dependent shift of the resonator frequency [Wallraff05]. In the dispersive limit, the qubit transition frequency ω_0 is far detuned from the resonance frequency of the cavity ω_r, ensuring $g/(\omega_0 - \omega_r) \ll 1$. Since the resonant coupling strength g is much smaller than the detuning $\Delta = \omega_0 - \omega_r$, no energy exchange between the two systems can take place. For small photon numbers, the detection of the electromagnetic field leaking out of the cavity forms a quantum non demolition measurement of the qubit state [Gambetta07, Gambetta08]. Combining a transmon type qubit with a Josephson bifurcation amplifier [Vijay09] led to a single shot read-out with a visibility of 94% [Mallet09].

Reading out the qubit state in the linear dispersive regime, which ensures a QND type measurement, in a single experimental realization with high fidelity is a challenging task. The current measurement setup does not implement a quantum limited amplifier and the noise added by the cold HEMT dominates over the very weak signal, as described in Sec. 3.2. A bigger signal would come to the expense of a reduced qubit coherence [Boissonneault08, Boissonneault09] and would lead to a complete breakdown of the dispersive approximation for amplitudes larger than n_{crit}. The qubit induced nonlinear cavity response at very high powers (> 1000 photons) is, however, providing an embedded bistable read-out device that can provide high fidelity single shot qubit measurements [Boissonneault10, Reed10a] and is discussed in more detail in chapter 6.

Another way to improve the signal to noise ratio, but not the single shot fidelity, demonstrated in the following, is to repeat each measurement many times, as described in Sec. 3.4, and then average.

4.1 Cavity-Bloch equations

To quantitatively study the dynamics of the coupled qubit-resonator system we derive Bloch-like equations of motions for the expectation value of the qubit operators $\langle \hat{\sigma}_i \rangle$ ($i = x, y, z$) and the resonator field operator $\langle \hat{a} \rangle$, valid in the limit of an ensemble average, realized for example by repeating many times the same experiment. Averaging over many identical experimental runs only provides information on the dynamics of an ensemble average, ruling out the observation of phenomena like quantum jumps.

The Hamiltonian (2.24) does not account for any external fields and needs to be extended to take into account the measurement tone applied to the resonator and for the coherent microwave field used to control the qubit state. The driving term of the Hamiltonian [Walls94]

$$H_{drive} = \hbar \left(\epsilon_m(t) \hat{a}^\dagger e^{-i\omega_m t} + \epsilon_s(t) \hat{\sigma}_+ e^{-i\omega_s t} + h.c. \right), \tag{4.3}$$

models a coherent measurement field applied to the cavity with amplitude $\epsilon_m(t)$ and frequency ω_m, and a second tone at frequency ω_s, with amplitude $\epsilon_s(t)$ driving the qubit over a gate-line, directly coupled via C_s to the qubit, see Sec. 3.1. The dispersive transformation, see Sec. 2.4, applied to the Hamiltonian (4.3), to first order in λ, generates an additional term $\lambda \epsilon_m(t) \hat{\sigma}_+ e^{-i\omega_m t} + h.c.$. This term describes the drive of qubit transitions over the resonator, with a reduced Rabi rate $(g/\Delta)\epsilon_m$.

To first order in λ in the dispersive approximation, the dynamics of the system in presence of dissipation and dephasing is described by a Lindblad-type master equation [Lindblad76]

$$\dot{\rho} = -\frac{i}{\hbar}[H^D, \rho] + \kappa \mathcal{D}[\hat{a}]\rho + \gamma_1 \mathcal{D}[\hat{\sigma}_-]\rho + \frac{\gamma_\phi}{2}\mathcal{D}[\hat{\sigma}_z]\rho \equiv \mathcal{L}\rho, \tag{4.4}$$

where $H^D = H^D_{JC} + H^D_{drive}$, H^D_{JC} is defined in Eq. (2.24) and $\mathcal{D}[\hat{A}]\rho = \hat{A}\rho\hat{A}^\dagger - \hat{A}^\dagger\hat{A}\rho/2 - \rho\hat{A}^\dagger\hat{A}/2$. Here, $\gamma_1 = 1/T_1$ is the qubit energy decay rate and γ_ϕ the qubit pure dephasing rate. These equations are valid in the Born-Markov approximation, where the reservoirs are assumed to be big, weakly coupled and memoryless or in other words induce white noise. At a small photon number $n \ll n_{ncrit} = |\Delta^2_{ar}|/4g^2$ and with γ_1 exceeding the Purcell decay rate [Houck08] we can neglect higher-order corrections [Boissonneault08, Boissonneault09]. For a full derivation up to third order one can refer to [Boissonneault07].

Equation (4.4) is not well suited for numerical solution since it contains oscillatory terms with high frequencies (of order of GHz). A transformation to a rotating frame with the unitary

CHAPTER 4. DISPERSIVE QUANTUM STATE READOUT

transformation

$$\mathbb{R} = e^{i(\omega_m \hat{a}^\dagger \hat{a} + \omega_s \hat{\sigma}_z/2)t} \quad (4.5)$$

can solve this technical difficulty. Assuming $\omega_r \approx \omega_m$ and $\omega_0 \approx \omega_s$, one can drop the fast oscillating terms in the so called rotating frame approximation and find

$$\begin{aligned} H_{RF}^D &= \Delta_{rm} \hat{a}^\dagger \hat{a} + \left[\Delta_{0s} + 2\chi\left(\hat{a}^\dagger \hat{a} + \frac{1}{2}\right)\right] \frac{\hat{\sigma}_z}{2} + \\ &\quad (\epsilon_m \hat{a}^\dagger + \epsilon_m^* \hat{a}) + [\Re(\epsilon_s)\hat{\sigma}_x - \Im(\epsilon_s)\hat{\sigma}_y], \end{aligned} \quad (4.6)$$

where we have defined $\Delta_{0s} = \omega_0 - \omega_s$ and $\Delta_{rm} = \omega_r - \omega_m$ as the detuning of the control and measurement microwave fields from the qubit and cavity frequency, respectively.

Combining this Hamiltonian with the master equation (4.4) leads to an infinite set of coupled equations for the expectation values. For instance, the differential equation for $\langle \hat{a} \rangle$ involves terms proportional to $\langle \hat{a}\hat{\sigma}_z \rangle$, $\langle \hat{a}^\dagger \hat{a}\hat{a}\hat{\sigma}_z \rangle$ and $\langle \hat{a}\hat{\sigma}_x \rangle$, which in turn involve even higher order terms. We therefore truncate this infinite series by factoring higher order terms $\langle \hat{a}^\dagger \hat{a}\hat{\sigma}_i \rangle \approx \langle \hat{a}^\dagger \hat{a} \rangle \langle \hat{\sigma}_i \rangle$ and $\langle \hat{a}^\dagger \hat{a}\hat{a}\hat{\sigma}_i \rangle \approx \langle \hat{a}^\dagger \hat{a} \rangle \langle \hat{a}\hat{\sigma}_i \rangle$, but keeping the terms $\langle \hat{a}\hat{\sigma}_i \rangle$ which ensures that the field contains information about the qubit state. This choice of factorization yields the correct average values for coherent and Fock states [Boissonneault07] and leads to a complete set of eight coupled differential equations

$$\begin{aligned} d_t \langle \hat{a} \rangle &= -i\Delta_{rm} \langle \hat{a} \rangle - i\chi \langle \hat{a}\hat{\sigma}_z \rangle - i\epsilon_m - \frac{\kappa}{2} \langle \hat{a} \rangle, \\ d_t \langle \hat{\sigma}_z \rangle &= \epsilon_s \langle \hat{\sigma}_y \rangle - \gamma_1 (1 + \langle \hat{\sigma}_z \rangle), \\ d_t \langle \hat{\sigma}_x \rangle &= -\left[\Delta_{0s} + 2\chi\left(\langle \hat{a}^\dagger \hat{a} \rangle + \frac{1}{2}\right)\right] \langle \hat{\sigma}_y \rangle \\ &\quad - \left(\frac{\gamma_1}{2} + \gamma_\phi\right) \langle \hat{\sigma}_x \rangle, \\ d_t \langle \hat{\sigma}_y \rangle &= \left[\Delta_{0s} + 2\chi\left(\langle \hat{a}^\dagger \hat{a} \rangle + \frac{1}{2}\right)\right] \langle \hat{\sigma}_x \rangle \\ &\quad - \left(\frac{\gamma_1}{2} + \gamma_\phi\right) \langle \hat{\sigma}_y \rangle - \epsilon_s \langle \hat{\sigma}_z \rangle, \\ d_t \langle \hat{a}\hat{\sigma}_z \rangle &= -i\Delta_{rm} \langle \hat{a}\hat{\sigma}_z \rangle - i\chi \langle \hat{a} \rangle + \epsilon_s \langle \hat{a}\hat{\sigma}_y \rangle \\ &\quad -i\epsilon_m \langle \hat{\sigma}_z \rangle - \gamma_1 \langle \hat{a} \rangle - \left(\gamma_1 + \frac{\kappa}{2}\right) \langle \hat{a}\hat{\sigma}_z \rangle, \\ d_t \langle \hat{a}\hat{\sigma}_x \rangle &= -i\Delta_{rm} \langle \hat{a}\hat{\sigma}_x \rangle - \left[\Delta_{0s} + 2\chi\left(\langle \hat{a}^\dagger \hat{a} \rangle + 1\right)\right] \langle \hat{a}\hat{\sigma}_y \rangle \\ &\quad -i\epsilon_m \langle \hat{\sigma}_x \rangle - \left(\frac{\gamma_1}{2} + \gamma_\phi + \frac{\kappa}{2}\right) \langle \hat{a}\hat{\sigma}_x \rangle, \end{aligned}$$

4.2 Continuous measurement

$$\begin{aligned}
d_t \langle \hat{a}\hat{\sigma}_y \rangle &= -i\Delta_{rm} \langle \hat{a}\hat{\sigma}_y \rangle + \left[\Delta_{0s} + 2\chi\left(\langle \hat{a}^\dagger \hat{a}\rangle + 1\right)\right]\langle \hat{a}\hat{\sigma}_x \rangle \\
&\quad -i\epsilon_m \langle \hat{\sigma}_y \rangle - \left(\frac{\gamma_1}{2} + \gamma_\phi + \frac{\kappa}{2}\right)\langle \hat{a}\hat{\sigma}_y \rangle \\
&\quad -\epsilon_s \langle \hat{a}\hat{\sigma}_z \rangle, \\
d_t \langle \hat{a}^\dagger \hat{a} \rangle &= -2\epsilon_m \Im \langle \hat{a} \rangle - \kappa \langle \hat{a}^\dagger \hat{a} \rangle,
\end{aligned} \qquad (4.7)$$

which we refer to as *cavity-Bloch equations*. While these equations are apparently more complex than Eq. (4.4), they can be analytically solved in some cases and are much faster to solve numerically. Note, that they do not include measurement-induced dephasing caused by photon shot-noise [Blais04, Gambetta06], because only the expectation value of $\hat{a}^\dagger \hat{a}$ is taken into account, and higher order moments are omitted.

The stationary state with a constant cavity drive is

$$\begin{aligned}
\langle \hat{a} \rangle &= \frac{-2i\epsilon_m}{\kappa + 2i(\Delta_{rm} - \chi)}, \\
\langle \hat{\sigma}_z \rangle &= -1, \\
\langle \hat{\sigma}_x \rangle &= 0, \\
\langle \hat{\sigma}_y \rangle &= 0,
\end{aligned} \qquad (4.8)$$

reproducing the expected Lorentzian cavity response centered around $\Delta_{rm} - \chi$ and full width at half maximum κ. For the case of vanishing cavity and qubit drive ($\epsilon_m = \epsilon_s = 0$) and arbitrary qubit initial conditions $\langle \hat{\sigma}_i \rangle(0)$ and $\langle \hat{a} \rangle(0) = 0$, the exact solution is [Boissonneault07]

$$\begin{aligned}
\langle \hat{a} \rangle &= 0, \\
\langle \hat{\sigma}_x \rangle(t) &= \left[\langle \hat{\sigma}_x \rangle(0)\cos(\omega' t) - \langle \hat{\sigma}_y \rangle(0)\sin(\omega' t)\right]e^{-(\gamma_\phi + \gamma_1/2)t}, \\
\langle \hat{\sigma}_y \rangle(t) &= \left[\langle \hat{\sigma}_y \rangle(0)\cos(\omega' t) + \langle \hat{\sigma}_x \rangle(0)\sin(\omega' t)\right]e^{-(\gamma_\phi + \gamma_1/2)t}, \\
\langle \hat{\sigma}_z \rangle(t) &= -1 + e^{-\gamma_1 t}\left[1 + \langle \hat{\sigma}_z \rangle(0)\right],
\end{aligned} \qquad (4.9)$$

where $\omega' = \Delta_{0s} + \chi$ is the Ramsey frequency and $\gamma_2 = \gamma_\phi + \gamma_1/2$ the phase decoherence rate.

4.2 Continuous measurement

To experimentally determine the state of the qubit, we probe the dynamics of the resonator-qubit system by measuring the resonator transmission. The transmitted tone is demodulated,

CHAPTER 4. DISPERSIVE QUANTUM STATE READOUT

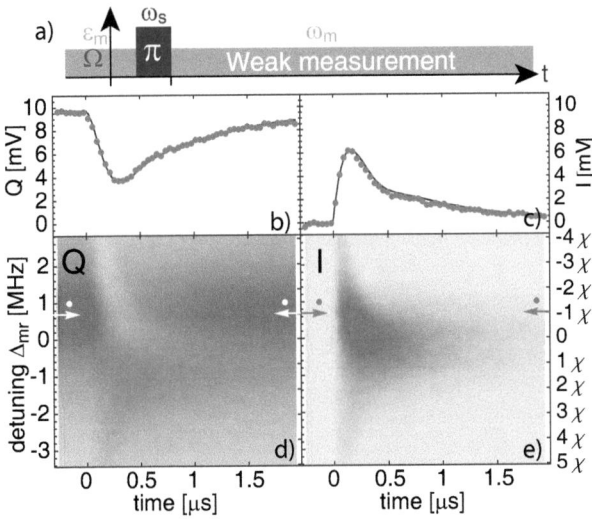

Figure 4.1 – a) Pulse scheme used to prepare and continuously read-out single qubit states. b) and c) averaged measurement response Q, I versus time t for a continuous weak measurement at the frequency $\omega_m = \omega_r - \chi$. Solid lines show the predicted response from the cavity-Bloch equations, Eq. (4.7). Time resolved data taken at different detunings Δ_{mr} is shown in d) and e). The arrows indicate the detuning at which the data shown in b) and c) is taken. The colormap codes red for a positive amplitude and white for zero.

digitized and averaged, see Sec. 3.4. When arbitrary qubit rotations can be performed, it is sufficient to consider the measurement response for the qubit prepared in either its ground $|g\rangle$ or excited state $|e\rangle$ for a full characterization of the qubit state [James01], a technique called state tomography. Figure 4.1a shows the pulse scheme used for the measurement. The time dependent quadrature amplitudes I and Q are measured at the cavity resonance frequency with the qubit in the ground state ($\omega_m = \omega_r - \chi$). The resonator is continuously driven at a measurement drive amplitude of $\epsilon_m^2 = \kappa/2$, populating the resonator with $\bar{n} \approx 1$ photons on average in resonance. A 10 ns long π pulse ending at time $t = 0$ and resonant with the ac-Stark [Schuster05] and Lamb shifted [Fragner08] qubit transition frequency $\omega_s = [\omega_0 + 2\chi(\langle a^\dagger a\rangle + 1/2)] := \omega_{s,res}$ is then applied to the qubit, see Figs. 4.1b and c. Qubit relaxation during the π pulse limits the achievable $|e\rangle$ state population to 99%, as obtained by solving the

4.2 Continuous measurement

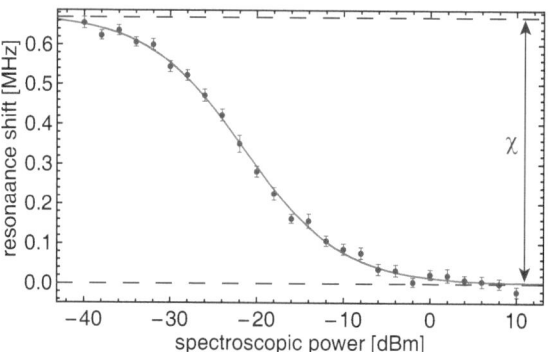

Figure 4.2 – Measured cavity resonance frequency shift versus resonant qubit driving power ϵ_s^2. The observed shift corresponds to the the cavity pull χ.

cavity-Bloch equations. For a similar discussion see [Chow09b]. This is within the statistical uncertainty of the detection. Furthermore, thermal excitation of the qubit is expected to be very low and is therefore neglected. The actual black body radiation in the cavity can be checked investigating the vacuum Rabi mode splitting [Fink10], finding a temperature smaller then 100 mK, or in the dispersive regime the qubit population can be inferred directly by spectroscopically characterizing the resonator response, as performed in Fig. 3.5.

The dependence of the quadrature components I and Q on the detuning Δ_{mr} of the measurement frequency from the bare resonator frequency is plotted in Fig. 4.1d and e. For clarity, the quadratures are rotated in the IQ-plane for each measurement frequency ω_m such that the Q quadrature is maximal in the steady-state (qubit in the ground state), resulting in $Q = A$ and $I = 0$ for $t \to \infty$. As a result, before the π-pulse the I quadrature is always zero.

The time and frequency dependence of the measurement signal is accurately described by the cavity-Bloch equations with a single set of independently measured, non adjustable parameters as indicated by the solid lines in Figs. 4.1b and c. The cavity resonance frequency is determined as $\omega_r/2\pi = 6.44252 \pm 0.00002$ GHz with a photon decay rate of $\kappa/2\pi = 1.69 \pm 0.02$ MHz. The qubit transition frequency is determined spectroscopically [Schreier08] as $\omega_0/2\pi \approx 4.009 \pm 0.001$ GHz with a charging energy of $E_c/h = 232.5 \pm 0.5$ MHz. The transition frequency is adjusted using external magnetic flux. The qubit-cavity coupling $g/2\pi = 134 \pm 1$ MHz is extracted from a measurement of the vacuum-Rabi mode splitting at $\omega_0 = \omega_r$ [Wallraff04].

CHAPTER 4. DISPERSIVE QUANTUM STATE READOUT

The cavity pull $\chi/2\pi = -0.69 \pm 0.02$ MHz is determined spectroscopically. This is done by measuring the cavity resonance frequency leaving the qubit in the ground state and then measuring its frequency shift applying a continuous coherent tone to the effective qubit transition frequency $\omega_{s,res}$, which saturates the qubit transition [Schuster05]. The resulting cavity resonance frequency versus qubit drive power is plotted in Fig. 4.2. The red line is only a guide to the eye used to extract the cavity pull using a fitting routine and not a calculation considering the full system response. When the qubit transition is saturated ($\epsilon_s \gg \gamma_1$), the resonator is shifted on average by χ. This value is in good agreement with the full transmon model taking into account higher levels [Koch07] ($\chi/2\pi = -0.71$ MHz).

In fitting the measurement response in Fig. 4.1, the qubit decay rate $\gamma_1/2\pi = 0.19 \pm 0.01$ MHz is used as an adjustable parameter which is equal to a measurement of γ_1 within the statistical uncertainty. In practice, the qubit decay rate is determined for one measured trace, and then kept fixed for all other traces. Note that, for short π-pulses, the dephasing rate γ_ϕ has no measurable influence on the solution of the equations. Additionally, a single scaling factor is introduced to relate the quadrature voltages at the output of the resonator to the digitized voltages after amplification.

To interpret the time and frequency dependence of the measurement signal shown in Fig. 4.1 it is instructive to plot I and Q as a function of ω_m at fixed times t, as shown in Fig. 4.3. With the qubit in $|g\rangle$, red points in Fig. 4.3, the resonator transmission exhibits the expected line shapes for both quadratures. When applying a π-pulse to prepare the qubit in $|e\rangle$, the cavity resonance frequency shifts by 2χ, but the transmitted quadratures respond only on a time scale corresponding to the photon lifetime $T_\kappa \equiv 1/\kappa$. The lineshape of the cavity transmission spectrum centered at $+\chi$ will only be reached in the limit of $T_1 \gg T_\kappa$, see dotted line in Fig. 4.3. The interplay of the cavity field rise time and the qubit decay time results in the observed dynamics of the cavity transmission in Figs. 4.1 and 4.3.

At time $t = 180$ ns $\sim 1.9\,T_\kappa \sim 0.2\,T_1$ after the preparation of $|e\rangle$ the shift of the cavity resonance to lower frequency towards $+\chi$ is clearly visible, see blue diamonds in Fig. 4.3. At $t = 740$ ns $\sim 7.9\,T_\kappa \sim 0.9\,T_1$, when $\approx 60\%$ of the excited state qubit population P_e is decayed, the measured curve is approximately the average between the steady state $|g\rangle$ and $|e\rangle$ responses, see green crosses in Fig. 4.3.

When looking at the time traces in Fig. 4.1d, the effective shift of the resonance to lower frequency explains the reduction of the signal in the Q quadrature for measurement detunings $\Delta_{mr} > -0.6$ MHz $\sim \chi$. For $\Delta_{mr} < -0.6$ MHz the amplitude is increased after the π-pulse

4.2 Continuous measurement

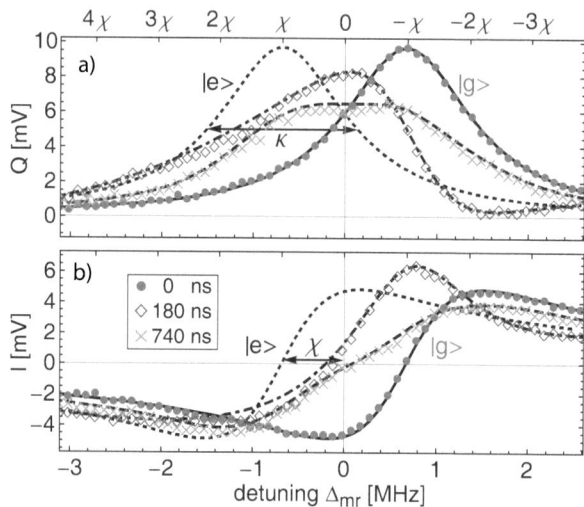

Figure 4.3 – Transmission spectrum of the resonator. The Q quadrature of the field is shown in a) and the I quadrature in b). The datapoints represent the instantaneous averaged response of the field before (red solid points), 180 ns (blue diamonds) and 740 ns after (green crosses) the π-pulse is applied. The underlying lines show the numerical simulations and the dotted line shows the expected response of the system for the qubit in the excited state $|e\rangle$ with infinite lifetime.

because the resonator is driven closer to resonance. Given our choice of the rotation of the traces in the IQ-plane, the I quadrature of Fig. 4.1e acts like a phase and always shows a positive response to the π-pulse.

The same considerations explain the features seen in the single measurement trace in Figs. 4.1b and c) taken at a measurement frequency corresponding to $\Delta_{mr} = -\chi$. The change of the I and Q quadratures on a timescale T_κ after the π-pulse reflects the relaxation of the field in response to the qubit excitation. The time scale of the return of the quadratures to their initial values is determined by the qubit decay at rate γ_1.

CHAPTER 4. DISPERSIVE QUANTUM STATE READOUT

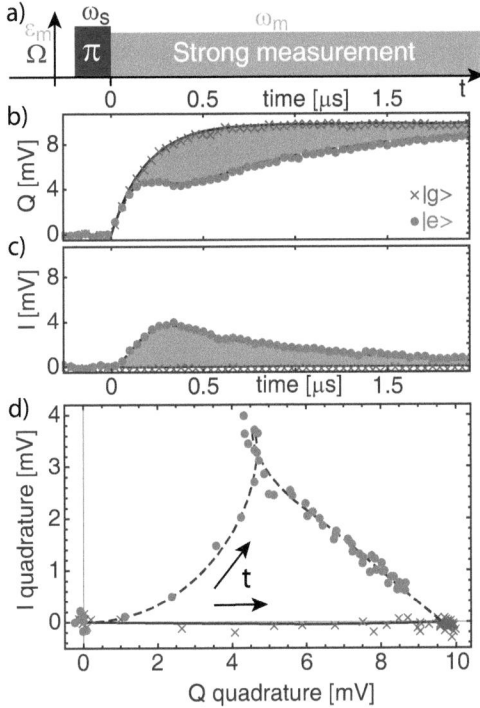

Figure 4.4 – a) Pulsed measurement scheme. b),c) Measurement response versus time for a pulsed averaged measurement taken for the same conditions as Fig. 4.1 for the Q and I quadrature respectively. Red points show the response for the qubit being prepared in the excited state $|e\rangle$. The blue crosses are the measured response to the qubit prepared in $|g\rangle$. The trajectory of the field in the IQ-plane is plotted in d), the arrows indicate the direction of the time.

4.3 Pulsed measurement

To avoid measurement-induced dephasing during the qubit manipulation most of the recent circuit QED experiments have been performed by probing the qubit state with pulsed measurements [Majer07, Schreier08, Chow09b, Leek09, Filipp09, DiCarlo09]. In contrast to a continuous measurement, the measurement tone is switched on only after the qubit state preparation is completed, see Fig. 4.4a for the pulse scheme. The absence of measurement photons during qubit manipulation also avoids the unwanted AC-Stark shift of the qubit transition frequency, thus simplifying qubit control.

With the qubit in $|g\rangle$ the resonator response reaches its steady state at the rate κ, which

4.3 Pulsed measurement

is seen in the exponential rise of the Q quadrature, see blue crosses in Fig. 4.4b. Since the resonator is measured on resonance at its pulled frequency $\Delta_{mr} = -\chi$, the I quadrature is left unchanged, see blue crosses in Fig. 4.4c. As in the continuous case, the resonator frequency is pulled to $\omega_r + \chi$ when the qubit is prepared in $|e\rangle$, see red dots in Figs. 4.4b and c. Since the resonator is now effectively driven off-resonantly, the transmitted signal has non vanishing I and Q quadrature components both of which contain information about the qubit state. With the measurement frequency still at $\Delta_{mr} = -\chi$, ringing occurs at the difference frequency $(\omega_r + \chi) - \omega_m = 2\chi$. At later times, the average response is approaching again the steady-state value as the qubit decays to $|g\rangle$ at the rate γ_1. As in the continuous case, the qubit lifetime in presence of measurement photons is obtained from a fit to the cavity-Bloch equations. Note, that the decay of the quadrature amplitudes shown in Figs. 4.4b and c does not directly correspond to the exponential decay of the qubit population $\langle \hat{\sigma}_z \rangle$, but is rather determined by the interplay of resonator and qubit evolution.

The dynamics of the I and Q quadrature amplitudes can also be represented in a phase-space plot, see Fig. 4.4d. The response for the qubit in $|g\rangle$ follows a straight line while the response for the qubit in $|e\rangle$ is more complex. The nontrivial shape of this curve reinforces the fact that both field quadratures contain information about the qubit state. It is obvious that a simple rotation in the IQ-plane cannot map the signal into a single quadrature.

Data taken at different measurement frequencies are shown in Fig. 4.5. As in Sec. 4.2, the I and Q components are rotated such that $Q = A$ and $I = 0$ in steady-state. For the theoretical curves (solid lines), the same set of parameters as for the analysis of the continuous measurement are used, leading to very good agreement. Figure 4.5a shows the measured response for the qubit in $|g\rangle$. The Q quadrature shows the expected exponential rise in the cavity population and for $t \gtrsim 0.5$ µs we recover the continuous measurement response. The I quadrature shows the transient part of the response during the initial population of the resonator, having a negative value (blue crosses) for measurements at a frequency above $\omega_r - \chi$ (blue detuned) and a positive value (red dots) at frequencies below $\omega_r - \chi$ (red detuned). Ringing can be observed when the measurement is off-resonant from the pulled cavity frequency. Figure 4.5b shows the response with the qubit prepared in $|e\rangle$. The response is similar to the one shown in Fig. 4.1 for the continuous measurement, if one omits the initial 100 ns where the resonator is populated.

CHAPTER 4. DISPERSIVE QUANTUM STATE READOUT

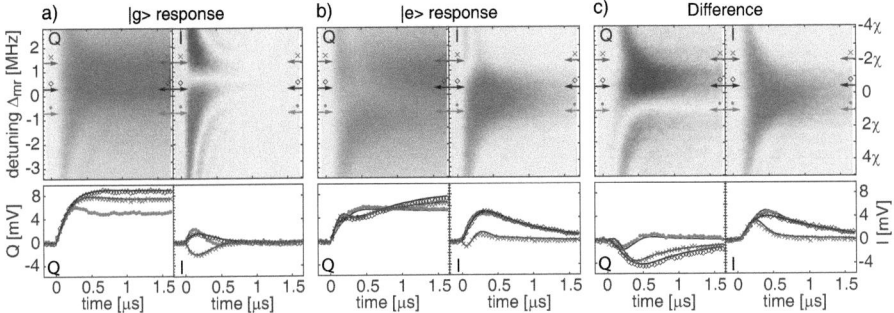

Figure 4.5 – I and Q quadratures for pulsed measurements at different detunings from the resonance frequency. The plots in a) are taken with the qubit in the ground state $|g\rangle$. b) displays the response of the system with the qubit prepared in the excited state $|e\rangle$. c) shows the result of the pointwise difference of the data acquired with the qubit in the ground state and the excited one. The lower panels show time traces taken at different detunings (blue crosses: $\Delta_{mr} = 1.4$ MHz, black diamonds: $\Delta_{mr} = 0.3$ MHz, red dots: $\Delta_{mr} = -0.7$ MHz) with comparison to theory (solid lines).

4.4 Population reconstruction

The detailed understanding of the dynamics of the dispersively coupled qubit/ resonator system can be used to infer the qubit excited state population $p_e = (\langle \hat{\sigma}_z \rangle + 1)/2$. Indeed, the difference in the measured response for a given unknown state $s_\rho(t)$ and the ground state response $s_g(t)$, which corresponds to the shaded area indicated in Fig. 4.4b and c, is directly proportional to p_e.

To explicitly state this relation, we introduce an effective qubit measurement operator $\hat{M}^i(t)$. Here, $i = I, Q$ denote the I and Q field quadratures used to measure the qubit state. In terms of this measurement operator, the I and Q components of the signal $s^i_\rho(t)$ for the qubit in state ρ before the measurement are given by

$$s^i_\rho(t) \equiv \langle \hat{M}^i(t) \rangle = \text{Tr}[\rho \hat{M}^i(t)], \qquad (4.10)$$

where $\hat{M}^i(t)$ is determined by the solution to the master equation (4.4). Analytical solutions

4.4 Population reconstruction

can be found in the limit of vanishing qubit decay [Filipp09],

$$\hat{M}^I(t) = \epsilon_m \frac{e^{-\kappa t/2}\left[2\hat{\chi}\cos(\hat{\chi}t) + \kappa\sin(\hat{\chi}t)\right] - 2\hat{\chi}}{\hat{\chi}^2 + (\kappa/2)^2},$$

$$\hat{M}^Q(t) = \epsilon_m \frac{e^{-\kappa t/2}\left[\kappa\cos(\hat{\chi}t) - 2\hat{\chi}\sin(\hat{\chi}t)\right] - \kappa}{\hat{\chi}^2 + (\kappa/2)^2}, \quad (4.11)$$

which depend on the operator $\hat{\chi} \equiv \Delta_{rm} + \chi\hat{\sigma}_z$ for the qubit state-dependent cavity pull. As a consequence of performing a quantum non-demolition measurement with only a few photons populating the resonator, mixing transitions between the two qubit states can be neglected [Blais04] and $\hat{M}^i(t)$ is diagonal at all times. The qubit then remains in an eigenstate during the measurement [Boissonneault09] and we can write $\hat{M}^i(t) = s_g^i(t)|g\rangle\langle g| + s_e^i(t)|e\rangle\langle e|$. The signals $s_g^i(t) = \text{Tr}[|g\rangle\langle g|\hat{M}^i(t)]$ and $s_e^i(t) = \text{Tr}[|e\rangle\langle e|\hat{M}^i(t)]$ are determined by Eq. (4.11) for the values $\langle\hat{\chi}\rangle = \Delta_{rm} \pm \chi$ corresponding to the qubit in the ground or excited state. To account for qubit relaxation, the cavity-Bloch equations (4.7) are solved to determine $s_{g/e}^i(t)$.

The qubit excited state population $p_e(\rho)$ in a given state ρ is determined by the normalized area between the measured signal s_ρ^i and theoretical ground state response s_g^i,

$$p_e(\rho) = \frac{1}{T}\sum_j \frac{s_\rho^i(t_j) - s_g^i(t_j)}{s_e^i(t_j) - s_g^i(t_j)}\Delta t, \quad (4.12)$$

where Δt denotes the discrete time step between datapoints. Replacing $s_\rho^i(t_j)$ with the corresponding expressions from Eq. (4.10), we notice that Eq. (4.12) simplifies to $p_e(\rho) = \text{Tr}[\rho|e\rangle\langle e|]$, demonstrating that the excited state population of an arbitrary input state is proportional to the normalized area between signal and ground state. Thus, the effective measurement operator $\hat{M}'^i = |e\rangle\langle e|$ defined by this procedure is equivalent to a projective measurement of the excited qubit state.

The measurement protocol can be summarized as follows: First, the relevant system parameters are determined in separate measurements. The qubit lifetime T_1, the single remaining parameter, is determined by applying a π-pulse to the qubit and analyzing the resulting transmitted signal. From this complete set of parameters, the signals $s_g^i(t)$ and $s_e^i(t)$ are computed. Finally, the excited state population p_e is calculated from the recorded signal $s_\rho^i(t)$ of an arbitrary qubit state ρ and the theoretical calculations of the responses $s_g^i(t)$ and $s_e^i(t)$, using Eq. (4.12), which amounts to a measurement of $\hat{M}'^i = |e\rangle\langle e|$. For the particular case of the

CHAPTER 4. DISPERSIVE QUANTUM STATE READOUT

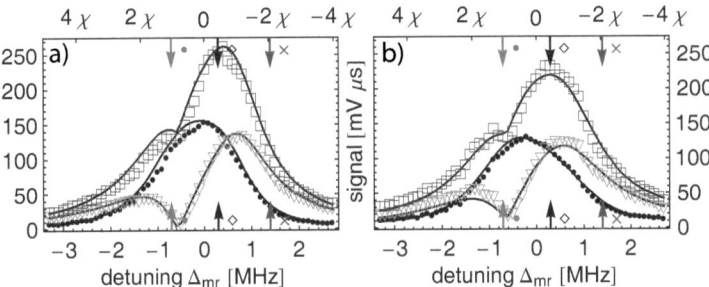

Figure 4.6 – Total integrated signal for the continuous a) and pulsed b) measurement, the triangles on the orange line represents the signal in the Q quadrature, while the squares on the magenta line show the I quadrature signal and the dots on the black line represents the sum of the two. The vertical lines indicate the position of the slices shown in Fig. 4.5.

qubit being in $|e\rangle$ after a π-pulse, the point-by-point difference signal is shown in Fig. 4.5c. Note, that the excited state population can also be directly inferred from a fit of the cavity-Bloch equations to $s_p^i(t)$ with p_e as free fit parameter. It is, however, computationally less intensive to calculate the population with the area method from Eq. (4.12), that is, to perform algebraic operations for the data analysis rather than employing a non-linear fit-routine. We have checked that both techniques provide the same results within the experimental precision.

This method has already been used in tomographic measurements to accurately measure both single and two-qubit density matrices [Filipp09, Leek09] and for the characterization of entangled states [DiCarlo09, Leek10].

4.4.1 Maximizing signal-to-noise ratio

The qubit state reconstruction procedure is valid for every chosen measurement frequency. For this reason we can maximize the total measurement signal

$$S^i(\Delta_{mr}) = \sum_j \left[s_e^i(\Delta_{mr}, t_j) - s_g^i(\Delta_{mr}, t_j) \right] \Delta t, \qquad (4.13)$$

used for the population reconstruction as a function of the detuning Δ_{mr}. $S^i(-\chi)$ corresponds to the red area shown in Fig. 4.4. The frequency maximizing the signal is dependent on the χ/κ ratio. For $\kappa \gg 2\chi$, the best measurement frequency is the bare resonance frequency, $\omega_m = \omega_r$, and the signal is then mainly in the phase component [Gambetta06]. In the opposite regime, where $2\chi \gg \kappa$, the resonator shifts enough to measure in resonance when the qubit is in the ground state $\omega_m = \omega_r - \chi$, and the dominant signal is in the amplitude of the transmitted signal.

For an intermediate regime, realized in the present experiment with $\chi/\kappa = 0.4$, there is a significant signal in both quadratures for every detuning which must be combined to optimize the measurement signal. The total signal $S^i(\Delta_{mr})$ acquired in 2 μs, versus measurement detuning is shown in Fig. 4.6, superimposed with the expected signal calculated using the cavity-Bloch equations. Since the population is linear in both quadratures, the simplest combination of the I and Q components to maximize $S^i(\Delta_{mr})$ is to add them together [Gambetta07]. The combined signal is plotted in Fig. 4.6 (black dots). The frequency maximizing $S^i(\Delta_{mr})$ for a continuous measurement is $\Delta_{mr} = 0.4$ MHz, while the best frequency for the pulsed measurement is $\Delta_{mr} = 0.3$ MHz. The system response at this detuning is shown in Fig. 4.5c (black trace), while the red trace indicates the position with almost no signal in the Q component.

4.5 Rabi oscillations measurements

To test the population reconstruction method experimentally, we perform a Rabi-oscillation experiment [Wallraff05], where a square pulse of variable length τ and amplitude ϵ_s is applied at the effective qubit transition frequency $\omega_{s,res}$. In the limit of large driving fields ($\epsilon_s \gg \gamma_1, \gamma_\phi$), the time-dependent population of the qubit can be approximated by the simplified expression [Allen87]

$$p_e(\tau) \cong \frac{1}{2} - \frac{1}{2} e^{-\frac{\tau}{4}(3\gamma_1 + 2\gamma_\phi)} \cos(\epsilon_s \tau / 2). \qquad (4.14)$$

Indeed, the population p_e obtained with the area method (Fig. 4.7a, points) has an rms deviation of less than 1% from the population predicted by Eq. (4.14), see the solid line in Fig. 4.7a.

In early experiments [Wallraff05], the population was reconstructed using the phase of the transmitted microwave instead of the quadrature amplitudes. Since the phase and the amplitude of an arbitrary signal are nonlinear functions of the I and Q quadratures, the measurement operator (4.11) is no more linear in the qubit population and Eq. (4.12) does not hold anymore. To demonstrate this effect, the qubit is prepared with different known populations and then measured with the area method using the transmitted phase and amplitude instead of the quadratures. The resulting estimated population versus known population is plotted in

CHAPTER 4. DISPERSIVE QUANTUM STATE READOUT

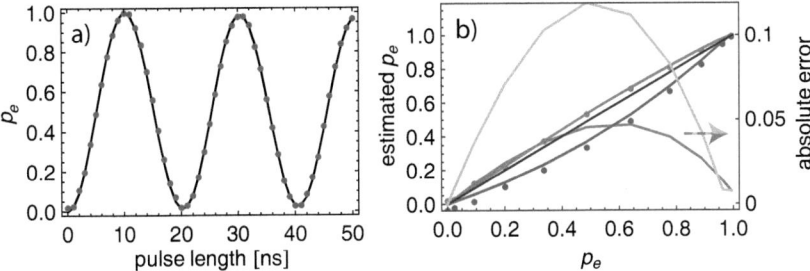

Figure 4.7 – a) Rabi oscillations of the qubit population p_e reconstructed by analyzing the pulsed Q response of the resonator (dots) using Eq. (4.12). The black line corresponds to the theoretical prediction calculated using cavity-Bloch equations with the following parameters: $\omega_a/2\pi = 4.504$ GHz, $\chi/2\pi = -1.02$ MHz, $T_1 = 860$ ns, $\epsilon_s/2\pi = 50$ MHz. b) Effective population versus population estimated with Eq. (4.12) for phase (red) and amplitude (blue) data. The magenta (amplitude) and cyan (phase) curves depict the absolute error in the estimated population on the right scala.

Fig. 4.7b, where the lines arise from a simulation using cavity-Bloch equations. The deviation from the ideal diagonal line in black is the absolute error in the population reconstruction and is shown on the same graph, on the right scala. For the parameters used in Fig. 4.7b the maximal error, at $p_e = 0.5$ is 0.05 and 0.12 for the amplitude and phase respectively, much bigger than the statistical uncertainty of 0.01. This systematic error depends on the chosen measurement frequency as well as the χ/κ-ratio. Performing a simulation with the parameters from [Wallraff05] leads to a maximal error of 0.02, which was not significant given their statistical uncertainty of 0.06.

4.6 Energy decoherence measurements

Improving the overall system coherence is a primary goal in current efforts in quantum information processing. The identification of loss mechanisms is a first step towards the realization of devices with enhanced coherence times. For example early transmon devices had limited lifetimes due to unwanted spontaneous decay trough the cavity. Engineering devices with smaller κ's or with new designs minimizing the Purcell effect [Reed10b] led to coherences up to a few μs in transmon type qubits [Houck08].

4.6 Energy decoherence measurements

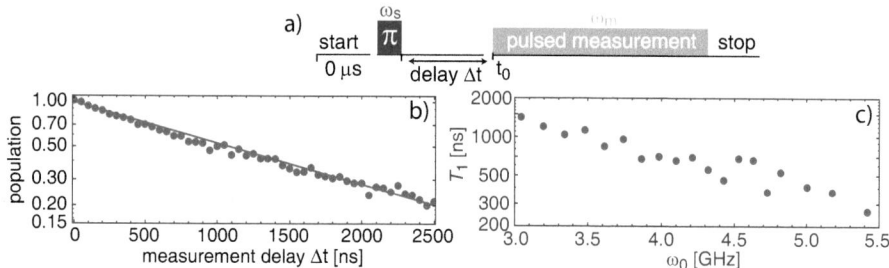

Figure 4.8 – a) Measurement scheme. b) Reconstructed qubit populations on a logarithmic scale versus measurement delay time δt after the preparation of the excited state. The line is the best exponential fit to the data with $T_1 = 1.2$ μs. c) Energy decay time versus qubit transition frequency.

Having demonstrated the ability to excite and read-out the qubit population with high accuracy we can implement a method to asses the energy relaxation time $T_1 = 1/\gamma_1$ of the qubit independently of the field inside the cavity. This is necessary because the rates fitted to the cavity-Bloch equations (4.7) can contain a significant contribution from photon dependent relaxation [Boissonneault08, Boissonneault09] due to the measurement photons. The determination of the energy relaxation rate is crucial to asses the quality of a given qubit implementation. The ratio of the energy decoherence time and the time needed for single and multi-qubit operations gives an upper limit to the number of operations that can be performed in an actual implementation of a quantum computer.

To determine the energy relaxation rate in absence of measurement photons, we prepare the qubit in the excited state, wait for the time Δt during which spontaneous decay may occur and then start the measurement to infer the population at the time t_0, as sketched in Fig. 4.8a. Fitting the decay shown in Fig. 4.8b to a rate equation results in a measurement of the systems energy relaxation, which is in good agreement with the value fitted to the cavity-Bloch equations for $n \ll n_{crit}$. The rates obtained from the two methods in a single experimental run usually deviate less then 10 %, which is of the same order as the fluctuations in time observed over several successive measurements.

Figure 4.8c shows the dependence of the qubit energy relaxation time on the qubit transition frequency ω_0. T_1 is not Purcell limited in this far detuned regime and shows an intrinsic limited $Q \approx 20'000$, similarly to [Houck08, Reed10b]. The decoherence source is under

CHAPTER 4. DISPERSIVE QUANTUM STATE READOUT

active investigation, dielectric loss in the tunnel barrier [Martinis05], surface two level systems [Shnirman05, O'Connell08] or nonequilibrium quasiparticles [Martinis09] have been proposed as possible sources.

4.7 T_2 measurements using Ramsey oscillations

The preservation of the excitation during the operation is not enough for quantum information tasks, but the qubit phase plays a central role [Childs10]. In general, one can identify two contributions to the phase decoherence rate $\gamma_2 = 1/T_2$: the decoherence related to the energy decay rate γ_1 and an additional decoherence rate arising from slow (on the timescale $1/\omega_0$) fluctuations of the qubit transition frequency, called pure dephasing γ_ϕ [Nielsen00]. The dephasing time T_2 describes the timescale at which the off diagonal terms of the qubits density matrix decay and is usually defined as

$$\rho_{01} = e^{i\omega_0 t} e^{-t/T_2}. \tag{4.15}$$

The environment model used to generate the cavity-Bloch equations makes similar assumptions about the noise coupling mechanism. This can be seen by comparing the exact solutions, Eq. (4.9) to Eq. (4.15), finding $\gamma_2 = \gamma_1/2 + \gamma_\phi$. In superconducting qubits, these fluctuations can arise from charge noise, flux noise and critical current noise [Vion02, Nakamura02, Koch07, Pashkin09].

To investigate the phase decoherence, the most direct way is to monitor the decay of the off diagonal terms of the density matrix for a state with a well defined phase such as $|\Psi_{ge}\rangle = (|g\rangle + |e\rangle)/\sqrt{2}$. This can be performed with state tomography or by mapping the phase information to a population that can be easily read-out. A simple implementation of this method, called Ramsey experiment, is to generate the state $|\Psi_{ge}\rangle$ with a $\pi/2$-pulse and then rotate the state in the Bloch-sphere after a free evolution time Δt with a second $\pi/2$-pulse, as sketched in Fig. 4.9a. The state evolution in the rotating frame can be approximated by

$$p_e = \frac{1}{2} + \frac{1}{2}\left[\cos(\omega' t) e^{-t/T_2}\right], \tag{4.16}$$

where $\omega' = \omega_0 - \omega_s + \chi$ denotes the detuning of the qubit drive. It is clear from Eq. (4.16), that this experiment is also useful to determine accurately the Lamb shifted qubit transition frequency. The method is robust against pulse imperfections which would deteriorate the

4.7 T_2 measurements using Ramsey oscillations

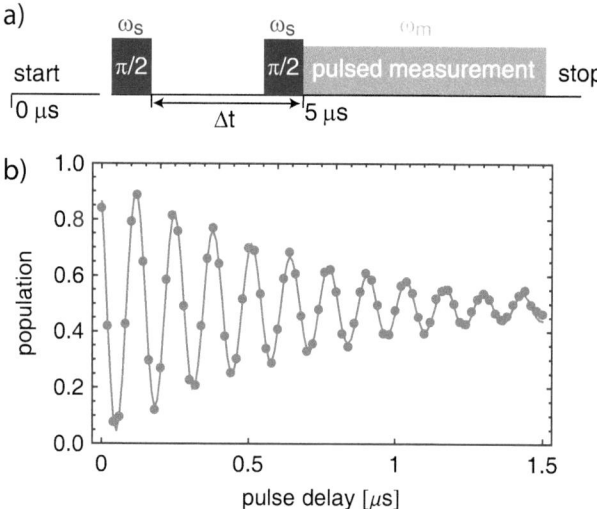

Figure 4.9 – a) Measurement scheme. b) Reconstructed qubit populations versus pulse delay time Δt after the preparation of the equal superposition state $|\Psi_{ge}\rangle$. The line is calculated with Bloch equations using $T_2 = 640$ ns and the actual pulse lengths of 12 ns.

visibility of the oscillations but not affect the frequency.

The results of such an experiment are shown in Fig. 4.9b, where coherent Ramsey oscillations at the chosen detuning frequency $\omega'/2\pi = 7.5$ MHz can be seen. The line is not a fit to Eq. (4.16) but a numerical simulation of the Bloch equations taking into account for the finite pulse lengths with a pure dephasing time $T_\phi = 1.1$ μs. The model is in very good agreement with the data, showing a maximal deviation of only 2%. A pure dephasing time of similar magnitude as T_1 is typical for transmon samples fabricated at ETH.

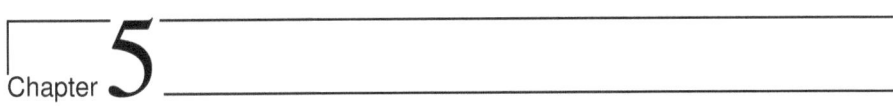

Generalization to 3-levels

Spin 1/2 or equivalent two-level systems are the most common computational primitive for quantum information processing [Nielsen00]. Using physical systems with higher dimensional Hilbert spaces instead of qubits has a number of potential advantages. They simplify quantum gates [Lanyon09], can naturally simulate physical systems with spin greater than 1/2 [Neeley09], improve security in quantum key distribution [Cerf02, Durt04] and show stronger violations of local realism when prepared in entangled states [Kaszlikowski00, Inoue09]. Multilevel systems have been successfully realized in photon orbital angular momentum states [Mair01, Molina-Terriza04], energy-time entangled qutrits [Thew04] and polarization states of multiple photons [Vallone07]. Multiple levels were used before for pump-probe read-out of superconducting phase qubits [Martinis02, Cooper04, Lucero08], were observed in the nonlinear scaling of the Rabi frequency of DC SQUID's [Murali04, Claudon04, Dutta08, Ferrón10] and were explicitly populated and used to emulate the dynamics of single spins [Neeley09]. In solid state devices, the experimental demonstration of full quantum state tomography [Thew02] of the generated states, i.e. a full characterization of the qutrit, is currently actively pursued by a number of groups.

In circuit QED, the third level has also been used, for instance, in a measurement of the Autler-Townes doublet in a pump-probe experiment [Baur09, Sillanpää09]. It has also been crucial in the realization of the first quantum algorithms in superconducting circuits [DiCarlo09] and is used in a number of recent quantum optical investigations, e.g. in Ref. [Abdumalikov10].

5.1 Three level Cavity-Bloch equations

To generalize the results of section 4.1 and study the system response to the excitation of more than two levels we consider the full dispersive Hamiltonian (2.33) instead of the two level approximation Eq. (2.24). Similar as in the two level case, the non-resonant interaction with the transmon in state $|n\rangle$ leads to a dispersive shift $S_n = -(\chi_n - \chi_{n-1})$ in the cavity frequency, different for each level. The expected transmission spectrum for the in-phase quadrature is sketched in Fig. 5.1a for different transmon states. Neglecting decoherence for now and choosing an appropriate measurement frequency ($\Delta_{rm} = \omega_r - \omega_m = 5.1$ MHz for example) gives a different signal α_n for each transmon state, allowing for a measurement of the population in each state.

Similarly to Eq. (4.4), we quantitatively study the dynamics of the coupled artificial atom-resonator system, taking into account for the system decoherence by performing an ensemble average. Each experiment is repeated many times to average out the experimental noise and to acquire enough statistics such that we can define the expectation values of the multilevel system projectors $\langle |i\rangle\langle i|\rangle$ ($i = 0, 1, 2$). The master equation (4.4) is then expanded to account for multilevel decoherence:

$$\dot{\rho} = -\frac{i}{\hbar}[H, \rho] + \kappa \mathcal{D}[\hat{a}]\rho + \sum_{i=1}^{n-1} \gamma_1^i \mathcal{D}[|i+1\rangle\langle i|]\rho + \sum_{i=1}^{n-1} \gamma_\phi^i \mathcal{D}[|i\rangle\langle i|]\rho, \qquad (5.1)$$

where γ_1^i and γ_ϕ^i are the energy decay rate and pure dephasing rate of level i, respectively. The Hamiltonian in the rotating frame

$$\begin{aligned}H^{RF} &= \Delta_{rm}\hat{a}^\dagger\hat{a} + \sum_{i=1}^{n}\Delta_{is}|i\rangle\langle i| + \sum_{i=0}^{n} S_i|i\rangle\langle i|\hat{a}^\dagger\hat{a} + \\ &+ \sum_{i=0}^{n-1}\epsilon_i\left(|i+1\rangle\langle i| + |i\rangle\langle i+1|\right) + \epsilon_m\left(\hat{a} + \hat{a}^\dagger\right),\end{aligned} \qquad (5.2)$$

takes into account two independent drives with amplitudes ϵ_i addressing the $0 \leftrightarrow 1$ and $1 \leftrightarrow 2$ transition respectively with a detuning of Δ_{is}. The infinite set of differential equations originating from the combination of Eq. (5.1) and Eq. (5.2) is factorized similarly to the two level cavity-Bloch equations, leading to 20 coupled differential equations which can be efficiently solved numerically, see appendix B.1.

For a first characterization of the read-out of higher levels, the transmon is prepared in one

CHAPTER 5. GENERALIZATION TO 3-LEVELS

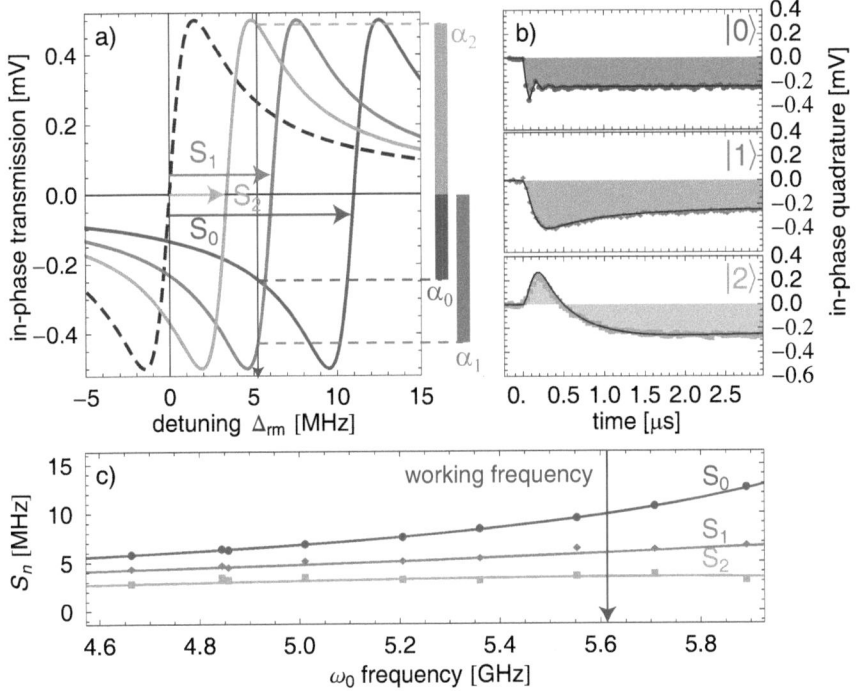

Figure 5.1 – a) Calculated in-phase transmission through the resonator for transmon states $n = 0, 1, 2$. The dashed curve indicates the bare resonator response. The vertical blue arrow indicates the detuning Δ_{rm} of the measurement tone. b) Pulsed I quadrature measurement responses for prepared states $|0\rangle$, $|1\rangle$ and $|2\rangle$. c) Measured dispersive shifts S_n versus transmon transition frequency ω_{01}. The solid lines are calculated within the linear dispersive approximation.

of its three lowest basis states $|i\rangle$ ($i = 0, 1, 2$). To perform this experiment we choose a detuning $\Delta_0 = \omega_0 - \omega_r = -1.319 \pm 0.001$ GHz. We extracted $g_0/2\pi = 115 \pm 1$ MHz from a measurement of the vacuum Rabi mode splitting [Wallraff04]. After state preparation, a coherent microwave tone is applied to the cavity and the state dependent transmission amplitude is measured, Fig. 5.1b. The amplitude of the tone was adjusted to maintain the average population of the cavity well below the critical photon number $n_{crit} = \Delta_0^2/4g_0^2 = 25$ [Blais04]. The time

5.1 Three level Cavity-Bloch equations

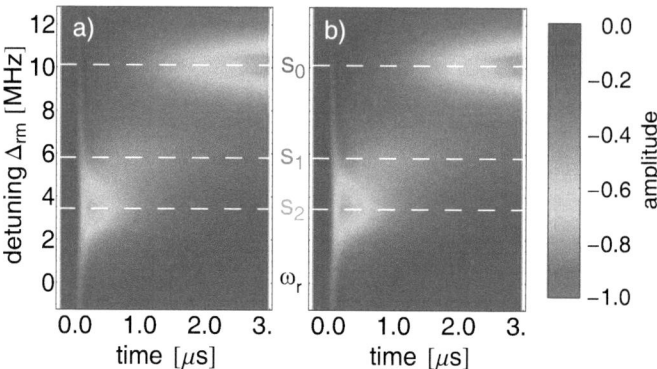

Figure 5.2 a) Measured Q quadrature of the resonator transmission versus time and measurement detuning for a preparation of the transmon in state $|2\rangle$. b) Calculation based on cavity-Bloch equations.

dependent transmission signals are characteristic for the prepared qubit states and agree well with the expected transmission calculated based on the extended cavity-Bloch equations. From the fits in Fig. 5.1b, we have extracted the state dependent cavity frequency shifts $S_{0,1,2}/2\pi = 10.0, 5.9, 3.4 \pm 0.1$ MHz, which are found to be within 0.1 MHz of the values calculated from independently measured Hamiltonian parameters. Also, the dispersive frequency shifts S_n measured in this way agree well with the linear dispersive model over a wide range of transmon transition frequencies ω_0, see Fig. 5.1c.

The frequency shifts can also be obtained by measuring the transmission amplitude over a wide range of detunings Δ_{rm} when preparing the transmon in the $|2\rangle$ state and observing its decay into the $|0\rangle$ state. Three distinct maxima in the measured Q quadrature, see Fig. 5.2a, located at the expected frequencies shifted by an amount S_n from ω_r are characteristic for the measurement of the $n = 0, 1, 2$ states of the transmon. The peaks appear successively in time, as the transmon sequentially decays from $|2\rangle$ to $|1\rangle$ to the ground state $|0\rangle$. Sequential decay is expected due to the near harmonicity of the transmon qubit, for which only nearest-neighbor transitions are important [Koch07].

The Q quadrature calculated from cavity-Bloch equations is in good agreement with the measurement data and yields the energy relaxation times of the first and second excited state $T_1^1 = 800 \pm 50$ ns and $T_1^2 = 700 \pm 50$ ns as the only fit parameters, see Fig. 5.2b. The relaxation

CHAPTER 5. GENERALIZATION TO 3-LEVELS

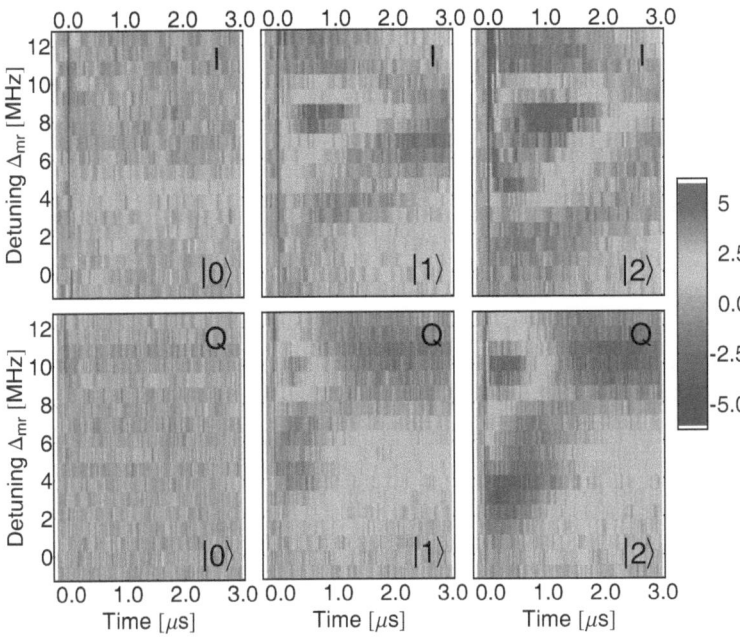

Figure 5.3 – Absolute difference between the measured and calculated I/Q quadrature response for prepared $|0\rangle$, $|1\rangle$ and $|2\rangle$ states. Red and blue indicate ± 5 % difference respectively, while light green indicates perfect agreement.

times are much longer than the typical time required to prepare the state and allow for a maximum second excited state population of 97%, limited by population decay during state preparation [Chow09b]. The absolute difference between data and theory, plotted in Fig. 5.3, is at most 3% at any given point indicating our ability to populate and measure the second excited state with high fidelity. The small difference could be ascribed to an imperfect state preparation due to our manipulation pulses or to an imprecise description of the coupled qutrit-cavity system. Equation (5.2), is only a first order approximation in λ and the factorization of terms, leading to the cavity-Bloch equations neglects higher-order corrections [Boissonneault08, Boissonneault09].

5.2 3-level population reconstruction

To realize high-fidelity qutrit control, arbitrary rotations in the three state Hilbert space with well defined phases and amplitudes are essential. Calibration of frequency, signal power and relative phases has to be performed based only on the population measurements of the qutrit states. To do so, we notice that the weak measurement partially projects the quantum state into one of its eigenstates $|0\rangle$, $|1\rangle$ or $|2\rangle$ in each preparation and measurement sequence [Blais04, Filipp09]. The average over many realizations of this sequence, which leads to the traces in Fig. 5.1b, can therefore be described as a weighted sum over the contributions of the different measured states. This suggests the possibility of simultaneously extracting the populations of all three levels from a single averaged time-resolved measurement trace. Formally, the projective quantum non-demolition measurement gives rise to the following operator, which is diagonal in the three-level basis and linear in the population of the different states at all times,

$$\hat{M}_{I/Q}(t) = s_0^{I/Q}(t)|0\rangle\langle 0| + s_1^{I/Q}(t)|1\rangle\langle 1| + s_2^{I/Q}(t)|2\rangle\langle 2|. \tag{5.3}$$

Here, $s_n^{I/Q}(t)$ are the averaged transmitted field amplitudes for the states $|n\rangle$ sketched in Fig 5.1a. The transmitted in-phase or quadrature amplitude

$$\langle I/Q_\rho(t)\rangle = \mathrm{Tr}\left[\hat{\rho}\hat{M}_{I/Q}(t)\right] = p_0 s_0^{I/Q}(t) + p_1 s_1^{I/Q}(t) + p_2 s_2^{I/Q}(t), \tag{5.4}$$

can be calculated for an arbitrary input state with density matrix ρ and populations p_i. Since any measured response is a linear combination of the known pure $|0\rangle$, $|1\rangle$ and $|2\rangle$ state responses weighted by p_i at each time step t_j, the populations can be reconstructed using an ordinary least squares linear regression analysis. In standard notation, Eq. (5.4) reads $y = X\beta + \epsilon$, where $y_j = A(t_j)$ are the measured quadrature amplitudes, $X_{ji} = s_i(t_j)$ the predicted pure state responses, $\beta_i = p_i$ and ϵ_j normally distributed noise with variance σ^2. The unknown populations β can be calculated by computing

$$\beta = \left(X^\mathrm{T} X\right)^{-1} X^\mathrm{T} y = X' y, \tag{5.5}$$

which can be performed very efficiently because one has to pseudoinvert X only once. The stored X' can then be reused for further population estimations.

5.3 Optimization of the measurement frequency

The reconstructed populations show larger statistical fluctuations than in the two-level case due to the pseudo-inversion of the ill-conditioned matrix X used to calculate the p_i. The condition is a property of the matrix X, influenced by the distinguishability between the different theoretically calculated traces, see Fig. 5.1b, and does not depend on the experimental noise. For an ill-conditioned matrix, a small error on the measured y, will result in a big error in the populations β. It is, however, an useful technique because it can reconstruct all qutrit state populations with a single measurement, enabling for example a Rabi or Ramsey experiment used to calibrate the tomographic measurement discussed in Sec. 5.9.

If the matrix $X^T X$ is nonsingular and the errors ϵ are normally distributed, the variance of the calculated populations β_i is well defined:

$$\sigma_i^2 = \frac{\epsilon^T \epsilon}{n-3} \left(X^T X \right)_{ii}^{-1}, \tag{5.6}$$

where n is the number of timesteps t_j. σ can be simulated for an arbitrary set of sample parameters by solving the three level cavity-Bloch equations, shown in Appendix B.1, and evaluating Eq. (5.6). Keeping all other experimental parameters constant, the errors can be minimized by optimizing the measurement detuning, see Fig. 5.4. The standard deviation of the reconstructed population of $|0\rangle$ (red trace), $|1\rangle$ (blue trace) and $|2\rangle$ (green trace), is shown versus measurement detuning. The parameters are taken form Sec. 5.1 and we assume a maximal signal of 1 V on resonance on top of a normally distributed noise background with a standard deviation of 0.01 V. Both quadratures are used to reconstruct the populations by calculating their weighted sum, with the weights given by their respective variances. The frequency dependence of the errors on the different states is nontrivial due to the complex interaction between the cavity field and the transmon decay which affects the distinguishability of the state responses. There is no global minimum, so one has to adapt the measurement detuning to the specific task. If one is interested only in the ground state population, the best detuning is S_0, while if one wants to minimize the errors for all populations one should choose 5.1 MHz, as shown in Fig. 5.1. These optimized measurement detunings are not globally valid and will change for different dispersive shifts S_i, cavity decoherence rates κ and qutrit decoherence rates γ_1^i.

For arbitrary system parameters it is not always clear which measurement frequency gen-

5.4 Rabi oscillations on the 2nd excited state

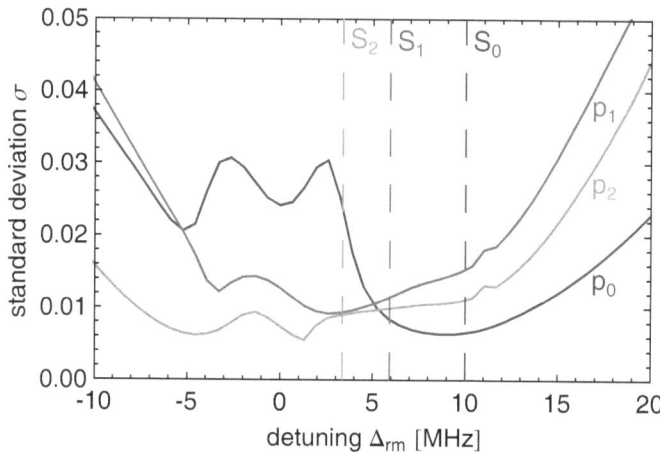

Figure 5.4 — Standard deviation of the reconstructed populations versus measurement detuning for the $|0\rangle$ (blue trace), $|1\rangle$ (red trace) and $|2\rangle$ (green trace) populations, simulated using the same parameters found in Sec. 5.1.

erates the set of data minimizing the statistical uncertainty of a desired qutrit population. Performing the analysis described in this section before the effective data is measured allows therefore for an optimal parameter choice, minimizing the experimental averaging time.

5.4 Rabi oscillations on the 2nd excited state

To demonstrate our ability to generate and read-out coherent populations of the $|2\rangle$ level we perform a simple time resolved Rabi oscillation experiment. This experiment is also routinely used to find the correct pulse amplitudes which generate π and $\pi/2$ rotations driving the transition between the first and second excited state. The implemented pulse scheme is depicted in Fig. 5.5a, where in contrast to Fig. 4.7 the transitions are driven by adjusting the pulse amplitudes instead of the pulse lengths. This is preferable at short pulse lengths of the order of 10 ns, because the employed commercial pulse generator (Tektronix AWG5014) has very good resolution in the amplitude (effectively 11 Bit) and not in the timing (1 ns timesteps).

CHAPTER 5. GENERALIZATION TO 3-LEVELS

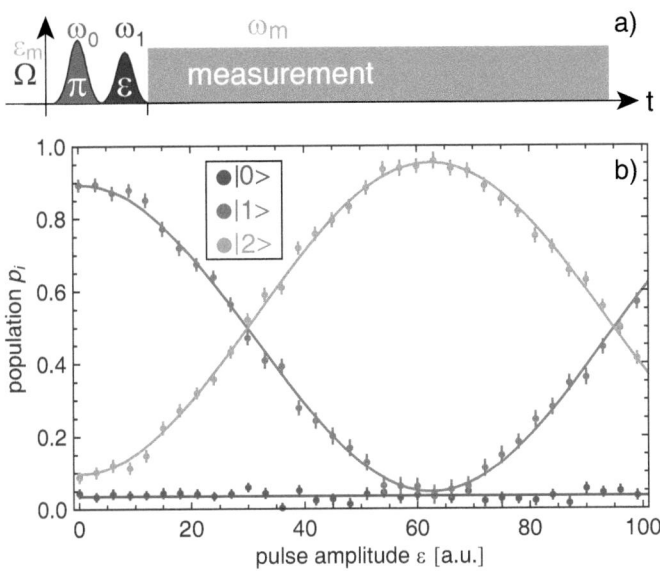

Figure 5.5 – a) Pulse scheme to perform Rabi oscillations between the $|1\rangle$ and $|2\rangle$ states. The qutrit is first prepared in the $|1\rangle$ state with a π pulse on the $0 \leftrightarrow 1$ transition followed by a pulse on the $1 \leftrightarrow 2$ transition with variable amplitude ϵ. b) Qutrit populations reconstructed with Eq. (5.5). The lines are fits to Eq. (4.14).

The reconstructed populations of all qutrit levels, using both measured quadrature amplitudes are shown in Fig. 5.5b. The evolution of the $|2\rangle$ population fits well to Eq. (4.14), from which the amplitude for a π pulse can be extracted, since the periodicity of the signal is not altered by small errors in the population reconstruction. There are, however, several imperfections such as a constant small (3%) population of the $|0\rangle$ level, and a reduced visibility of the Rabi oscillations, apparent in the offset in the initial population of the $|1\rangle$ (8%) and $|2\rangle$ (10%) states. There are two possible contributions to these imperfections: imperfect pulses generating imperfect final states or errors in the state read-out. The generation of improved pulses is described in Sec. 5.6, while errors in the state read-out could arise from small deviations from the linear dispersive Hamiltonian (4.6) or from slightly wrong parameters used to generate the X matrix from the cavity-Bloch equations. These errors are enhanced by the

79

same mechanism described in section 5.3 but are of systematic rather then statistic nature. To reduce the read-out errors, quantum state tomography [Thew02] can be implemented at the expense of more measurements and under the stringent requirement of accurate qutrit state manipulation, making the implementation of optimally controlled pulses unavoidable.

5.5 Measurement of the Rabi rates

Performing Rabi oscillations allows for a direct measurement of the coupling strengths g_i because the rate of the oscillation between the ground and excited state is proportional to the qubit dipole moment which in turn implies the coupling coefficients g_i. A direct measurement of g_i was done before on resonance [Fink08, Fink09], where $\omega_r \approx \omega_0$. It is, however, interesting to check the model also in the dispersive regime, where the transmon couples to an external drive field via the Hamiltonian [Blais04, Koch07]

$$H_{drive} = \hbar \Omega_i \left(|i\rangle\langle i+1| + |i+1\rangle\langle i| \right), \tag{5.7}$$

where the Rabi rate $\Omega_i = d_i E/\hbar$ is proportional to the electric field E generated by the drive tone and the transmon dipole moment $d_i = \hbar g_i / \mathcal{E}_{rms}$. The transmon is driven over a charge line, coupling directly to the qutrit, as sketched in Fig. 3.1, so that the microwave tone is not filtered by the cavity. When driven on resonance with the qubit, $\omega_s = \omega_i + L_i$, Rabi oscillations with frequency Ω_i are induced. A measurement of Ω_i at fixed drive amplitude E, allows therefore to determine g_i. According to equations (2.31) and (2.32), the matrix element part of the coupling strength ratio is expected to scale as $g_n/g_{n-1} \approx \sqrt{n+1}$ due to the almost harmonic oscillator wavefunction of the transmon qubit.

The measured Rabi frequencies for the first two transitions are observed to decrease for decreasing driving frequency ω_i, see Fig. 5.6. The observed behavior can be explained by the frequency dependent loss of the cables, which effectively changes the drive amplitude at the qutrit location as a function of frequency, as discussed in Sec. 3.1. The dashed lines in Fig. 5.6 depict the expected room temperature frequency dependent attenuation of the coaxial cables listed in Tab. 3.1 and are found to be in good agreement with the measured Rabi rate scaling.

To eliminate the uncertain attenuation in the cryostat, the measured Rabi frequencies on the different transitions taken at the same frequency are compared by taking Ω_1/Ω_0, as shown in the black crosses of Fig. 5.6, which gives 1.43 ± 0.04 on average, as expected.

CHAPTER 5. GENERALIZATION TO 3-LEVELS

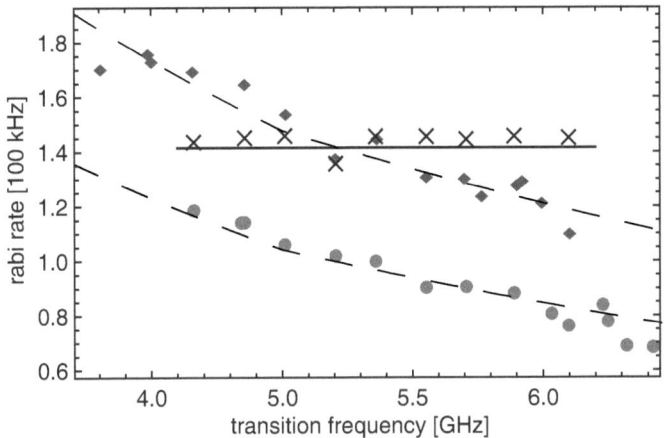

Figure 5.6 – Measured Rabi frequency on the $0 \leftrightarrow 1$ (red dots) and $1 \leftrightarrow 2$ (blue diamonds) transition versus transition frequency ω_0 and ω_1 respectively. The dashed lines show the calculated frequency dependent cable attenuation. g_1/g_0 at a given frequency and fixed drive amplitude is displayed as black crosses, while the black line indicates the number $\sqrt{2}$.

5.6 Pulse optimization

For a system with more than two levels and limited anharmonicity, the simple square pulses used in the previous sections do not provide a single qubit gate with high fidelity. If short square or Gauss pulses are used on the $0 \leftrightarrow 1$ transition, the second excited state is significantly populated. The mere presence of the second level also introduces an AC-Stark shift which results in phase errors on the first excited state. It is therefore unfeasible to prepare 3-level states with high fidelity relying on short square pulses. To avoid these problems we use optimal control techniques (see for example [Steffen03, Jirari05, Rebentrost09, Safaei09] for earlier results in superconducting circuits) to shape our pulses, implementing the Derivative Removal by Adiabatic Gate (DRAG) method proposed in [Motzoi09] and realized in [Chow09a]. There, the interaction of the driving Hamiltonian (4.3) with the first three transmon levels is considered

5.6 Pulse optimization

in the rotating frame

$$H_{drive} = \begin{pmatrix} 0 & \dfrac{\epsilon_x(t) + i\epsilon_y(t)}{2} & 0 \\ \dfrac{\epsilon_x(t) - i\epsilon_y(t)}{2} & -\dot\phi(t) & \sqrt{2}\dfrac{\epsilon_x(t) + i\epsilon_y(t)}{2} \\ 0 & \sqrt{2}\dfrac{\epsilon_x(t) - i\epsilon_y(t)}{2} & \Delta_{01} - 2\dot\phi(t), \end{pmatrix} \quad (5.8)$$

where ϵ_x and ϵ_y are the amplitudes of the two quadratures of the drive $\epsilon(t) = \epsilon_x(t)\cos(\omega_s t + \phi(t)) + \epsilon_y(t)\sin(\omega_s t + \phi(t))$ with time dependent phase $\phi(t)$. $\Delta_{01} = \omega_0 - \omega_1$ is the anharmonicity. This expression is different from the one derived in [Motzoi09], because it is technically easier to add an extra phase ϕ to the control pulses than to vary the qubit frequency accurately on this timescale. Changing the phase of the driving pulse equates to changing its frequency which is formally equivalent to a time dependent qubit transition frequency. One can use the free parameter $\epsilon_y(t)$, i. e. the second quadrature, and the phase $\phi(t)$ to eliminate the phase error and the leakage to the third level. The conditions found by transforming the Hamiltonian (5.8) adiabatically to the two dimensional qubit subspace are

$$\epsilon_y(t) = -\frac{\dot\epsilon_x(t)}{\Delta_{01}} \quad \text{and} \quad \phi(t) = \int_0^t \frac{3\epsilon_x^2(s)}{4\Delta_{01}} ds. \quad (5.9)$$

The leakage is eliminated to order $\epsilon_x^4/\Delta_{01}^3$ at each time and the proposed pulses do not require any sharp features if the intended pulse for the qubit state manipulation ϵ_x is a smooth function of time, such as a Gaussian, as shown in Fig. 5.7.

To experimentally test the feasibility of this method we choose a set of pulse sequences which are particularly prone to phase errors and vary the amplitude of the compensation quadrature ϵ_y. We implement Gauss pulses on ϵ_x of standard deviation 3 ns and total length 12 ns and vary the amplitude to implement π and $\pi/2$ rotations around the x-axis in the Bloch sphere. In Fig. 5.8a, the measured excited state population is plotted versus a linear scaling factor on ϵ_y for a $\pi/2$ rotation on the x-axis followed by a π rotation on the y-axis (blue dots). An ϵ_y scale of 1 results in the pulse shown in Fig. 5.7, defined by Eq. (5.9), while ϵ_y scale=0 implies $Q = 0$. As control experiments the second pulse is applied about the same axis as the first (red dots) and about the minus y-axis (black dots). For perfect pulses, all three sequences should result in the same population of 1/2 (neglecting decoherence during the pulses) while any error in the phase of the first pulse is transformed into a population error by the successive rotation

CHAPTER 5. GENERALIZATION TO 3-LEVELS

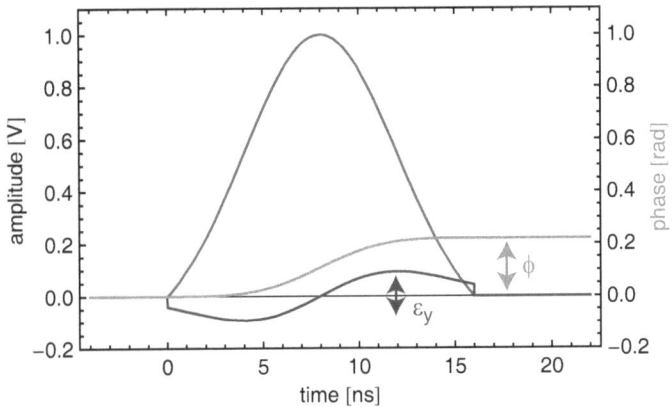

Figure 5.7 — Example of a DRAG pulse before it is upconverted to the qubit transition frequency. A gaussian pulse with amplitude $\epsilon_x(t)$ is generated on the I quadrature (red), while the $Q = \epsilon_y(t)$ quadrature is plotted in blue. The additional phase is shown in green, while the scaling parameters used for the pulse optimization are depicted as ϵ_y and ϕ. The Gaussian envelope has a standard deviation $\sigma = 4$ ns and the qubit has an anharmonicity $\Delta_{01}/2\pi = -300$ MHz.

on the other axis. This behavior is observed in the data, where phase errors are revealed by changes in the populations with opposite signs for rotations on the y- and minus y-axes, while the population remains unchanged in the control experiment for subsequent rotations on the x-axis. The experimental data confirms the validity of the approach finding the correct population of 0.5 for an ϵ_y scaling factor of 1, demonstrating the benefit of DRAG pulses versus a bare Gaussian which is implemented for a scaling factor of 0, where the error is as big as 20% in population after only two pulses.

The same experiment is performed by scaling linearly the phase compensation ϕ calculated in Eq. (5.9), and depicted in Fig. 5.7 using the correct quadrature compensation ϵ_y. The measured data is plotted in Fig. 5.8b, showing the same features discussed for Fig. 5.8a. The same considerations discussed above apply here, demonstrating the necessity of the phase compensation to get high pulse fidelities.

To further demonstrate the quality of the implemented pulses we measure a set of 25 different pulse sequences, chosen to perform rotations about all possible combinations of rotations about two different axis which would make any phase error measurable as a population error,

5.6 Pulse optimization

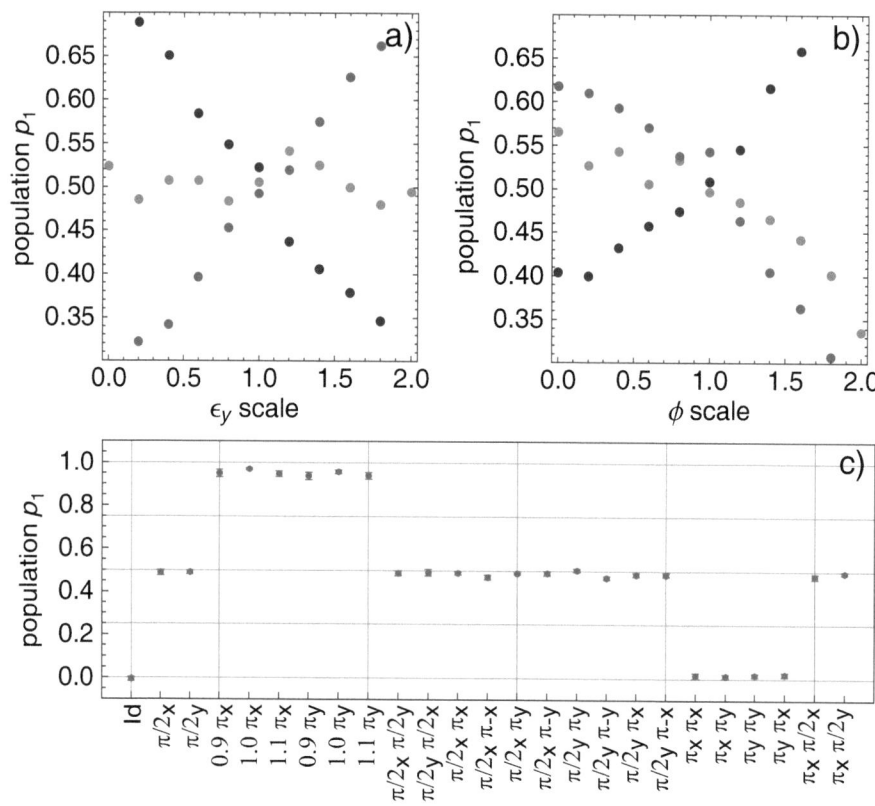

Figure 5.8 – Calibration of the DRAG pulses. a) Population versus scaling factor of the derivative component ϵ_y. b) Population versus scaling factor of the phase compensation factor ϕ. c) Benchmarking of the DRAG pulses with a set of standard qubit manipulations.

as shown in Fig 5.8c. The measured data does not show any systematic deviation from the expected populations for different pulse sequences, demonstrating the quality of the implemented pulses. To quantitatively asses remaining gate errors one could implement randomized gate benchmarking [Chow09b]. The small deviations of the single π pulse sequences from unity population are probably due to errors in the population reconstruction routine and are not further

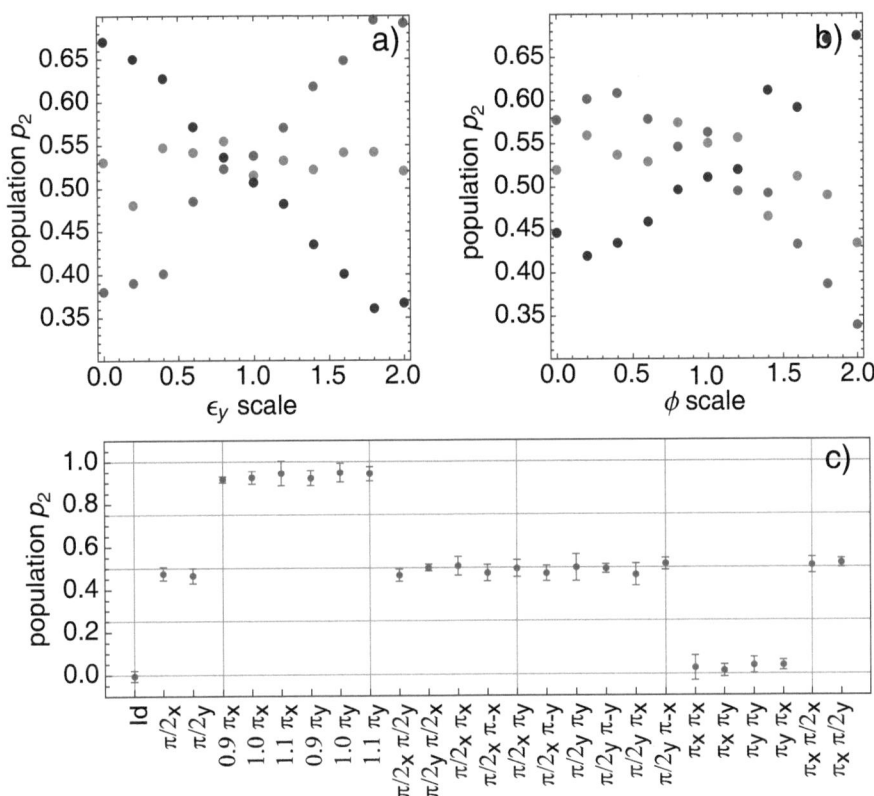

Figure 5.9 – Calibration of the DRAG pulses on the $1 \leftrightarrow 2$ transition. a) Population p_2 versus scaling factor of the derivative component ϵ_y. b) Population p_2 versus scaling factor of the phase compensation factor ϕ. c) Benchmarking of the DRAG pulses with a set of standard qutrit manipulations after the generation of the pure $|1\rangle$ state.

investigated in this section which focuses on the accurate control of the pulse phases.

We extend this technique to three levels using quadrature compensation and time-dependent phase ramps to suppress population leakage to other states and to obtain well defined phases. In theory, the same pulses as the ones described above should enable high fidelity transitions between higher levels [Motzoi]. To benchmark them, we first excite the qutrit to the $|1\rangle$ state with a DRAG π-pulse and then apply the pulse sequence described in the previous paragraph but on the transition between $|1\rangle$ and $|2\rangle$, using the same compensation factors derived in Eq. (5.9). Fig 5.9a and b show the same features as before, demonstrating our ability to perform rotations between the $|1\rangle$ and $|2\rangle$ states with high fidelity by avoiding phase errors. The population resulting from uncorrected pulses (with errors of order of 10% in population), shown in Fig. 5.9 with $\epsilon_y = 0$ or with $\Phi = 0$, clearly demonstrate how necessary the DRAG pulses are to ensure the correct gate operations. The fidelity of the pulses will be quantified in Sec. 5.9, where the states are analyzed using full quantum state tomography, but it is already clear from Fig. 5.9 that any remaining errors should be smaller then a few %.

5.7 T_1-time of the 2nd excited state

In Sec. 4.6, the energy decoherence rate of the first excited state is assessed, finding an intrinsically limited quality factor in the far detuned regime. Having demonstrated the ability of reading out the population of the second excited state, see Sec. 5.2, the question arises what the coherence time is for the second excited state and how it scales with detuning.

The decoherence rate γ_1^{12} fitted to the data shown for example in Fig. 5.2 to the extended cavity-Bloch equations shown in Appendix B.1 for the second excited state contains a larger contribution from photon induced relaxation than the first level because of the larger coupling coefficient [Boissonneault08, Boissonneault09, Boissonneault10]. A measurement of T_1^{12} using the delayed measurement procedure presented in Sec. 4.6 is used to investigate the energy relaxation rate of the second excited state in absence of measurement photons. Furthermore, a direct decay to the ground level T_1^{02} is suppressed [Koch07] and therefore not taken into account for in the three level cavity-Bloch equations. This must be verified experimentally.

The measurement, taken under the same conditions as described in Sec. 5.1, using the same sample, is shown in Fig. 5.10. The data is fitted to the rate equations

$$\begin{aligned} p_2'(t) &= -\gamma_1^{12} p_2(t) - \gamma_1^{02} p_2(t), \\ p_1'(t) &= -\gamma_1^{01} p_1(t) + \gamma_1^{12} p_2(t), \\ p_0'(t) &= \gamma_1^{02} p_2(t) + \gamma_1^{01} p_1(t), \end{aligned} \qquad (5.10)$$

finding $T_1^{12} = 1.08 \pm 0.05$ μs as only free parameter since $T_1^{01} = 1.13 \pm 0.05$ μs is measured independently in a previous experiment, using the methods presented in Sec. 4.6, and T_1^{02} is

CHAPTER 5. GENERALIZATION TO 3-LEVELS

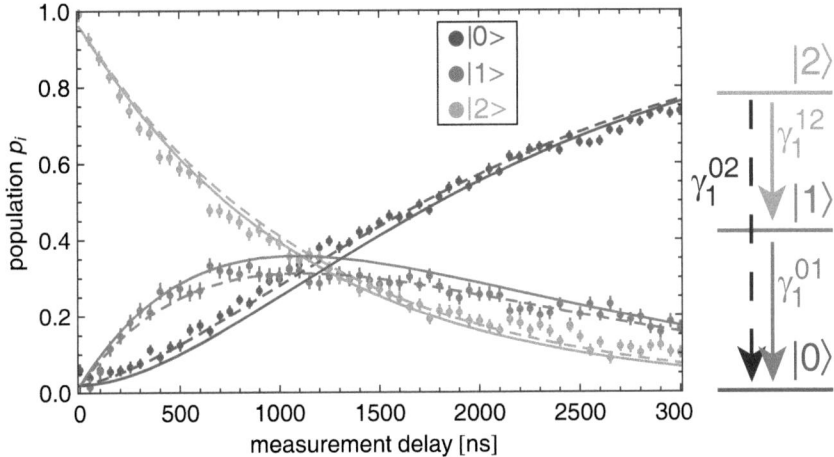

Figure 5.10 – Decay of the prepared pure $|2\rangle$ state. The $|2\rangle$ state is generated with two subsequent π-pulses on the ω_0 and ω_1 transitions and the measurement is delayed by a variable amount of time. The continuous solid lines result from a simple rate equation fit and the dashed line has T_1^{02} as additional free parameter. The relevant energy levels and decay rates are sketched on the right.

assumed to be infinite. Considering also direct decay from state $|2\rangle$ to $|0\rangle$, with coherence time T_1^{02}, the fit is slightly better (see dashed lines in Fig. 5.10, where the residues are smaller) and results in $T_1^{12} = 1.31 \pm 0.05$ µs and $T_1^{02} = 11 \pm 2$ µs. The accuracy of these coherence times is, however, low. This is apparent in the predicted difference of the population between the two models which is smaller than 5%. The rate γ_1^{02} is therefore sensitive on population differences in the percent range which could be ascribed to population reconstruction errors, which in turn, if taken into account in a new fit, would imply a different T_1^{02}. Thus, the found value for T_1^{02} should be considered as a lower bound, confirming the theoretical predictions.

A first comparison between the energy relaxation time of the first and second exited states, extracted from a fit to the cavity-Bloch equations over a range of frequencies below the resonator frequency ω_r results in a ratio of energy relaxation times of $T_1^{12}/T_1^{01} = 0.7 \pm 0.1$. There are, however, big fluctuations and the influence of the chosen constant measurement power over different detunings which imply different critical photon numbers n_{crit} and in turn different photon dependent decoherence rates has not been taken into account. This preliminary result, which has to be confirmed, is in agreement with the simple model of a noisy reservoir having a white spectrum (for example fluctuating two-level systems (TLS) with electric-dipole moments in the substrate [Constantin09]) coupling via g_i to each transition, predicting a scaling of the coherence times with approximately $1/\sqrt{2}$. In a different model, considering the transmon as nearly harmonic, the coherence time of the second excited state would be half the value of the first excited state, as found in phase qubits [Neeley09].

5.8 Phase coherence of the 2nd excited state

To asses the precise value of ω_1 and T_2^{12}, we perform a Ramsey experiment between the $|1\rangle$ and the $|2\rangle$ level, see Fig. 5.5. We apply a π-pulse at ω_0, to prepare the first excited state and then vary the delay time between two successive $\pi/2$ pulses applied at $\omega_1 - 7.5$ MHz and finally perform a pulsed qubit state read-out. Similarly to the simple two-level case, discussed in Sec. 4.7, the coherent oscillations allow to determine the transition frequency with an accuracy exceeding 100 kHz (which is a typical fluctuation rate of the qubit transition). To avoid any phase error, the DRAG pulses, presented in Sec. 5.6, must be driven on resonance.

The observed oscillatory decay in the qubit population with delay time Δt of the Ramsey experiment can be simulated with a Bloch equation from which the phase coherence of the second excited state is inferred. The energy decay rates found in Sec. 5.7 and the dephasing of the $|1\rangle$ state are kept fixed while $T_\phi^{12} = 650 \pm 50$ ns is extracted from the fit. Due to the decay of the $|1\rangle$ state, the $|2\rangle$ state population does not converge to 0.5 for long pulse delays and the ground state population monotonically grows to 1 because the last $\pi/2$-pulse, resonant on the $1 \leftrightarrow 2$ transition, has no effect on a pure $|0\rangle$ state. The good agreement with the theoretical prediction demonstrates the high phase coherence of the $|2\rangle$ level. The small systematic deviations observed in the data are due to the imperfect state read-out rather then imperfect state preparation. This can be seen confronting the data presented in Fig. 5.11 with the data shown in Fig. 5.13, taken in identical conditions but read-out using a tomographic method, described in the following section.

CHAPTER 5. GENERALIZATION TO 3-LEVELS

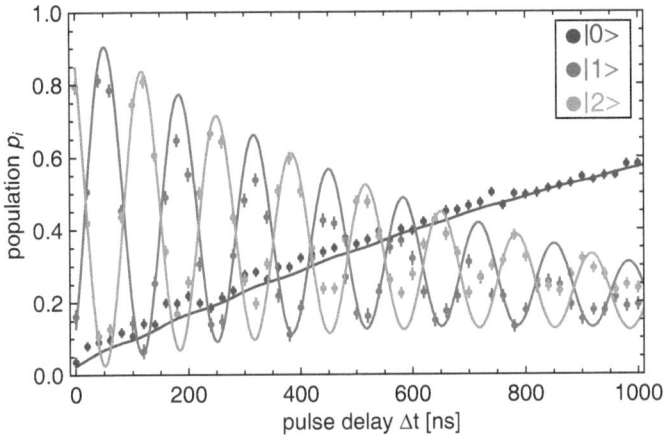

Figure 5.11 – Ramsey experiment between the $|1\rangle$ and $|2\rangle$ state. Initially, the state $|1\rangle$ is prepared and then the pulse scheme sketched in Fig. 4.9a is applied. The shown qutrit populations are reconstructed with Eq. (5.5) and the solid lines are a fit to three level Bloch equations with dephasing time $T_2^2 = 500 \pm 50$ ns.

5.9 3-level tomography

Using a measurement of the population alone does not provide any information on the phases of a given state. Using quantum state tomography [Thew02], the full density matrix of the first three levels of a transmon can be reconstructed. This is achieved by performing a complete set of nine independent measurements after preparation of a given state and calculating the density matrix based on the measurement outcomes. Since the measurement basis is fixed by the Hamiltonian (2.30), the state is rotated by applying the following pulses prior to measurement

$$\mathbb{I}, \ \left(\frac{\pi}{2}\right)_x^{01}, \ \left(\frac{\pi}{2}\right)_y^{01}, \ (\pi)_x^{01}, \ \left(\frac{\pi}{2}\right)_x^{12}, \ \left(\frac{\pi}{2}\right)_y^{12},$$
$$(\pi)_x^{01} \left(\frac{\pi}{2}\right)_x^{12}, \ (\pi)_x^{01} \left(\frac{\pi}{2}\right)_y^{12}, \ (\pi)_x^{01} (\pi)_x^{12}, \quad (5.11)$$

where $\mathbb{I}$ denotes the identity and $(\theta)_a^{ij}$ denotes a pulse resulting in a state rotation of angle θ on the ij-transition about the a-axis. For each of these unitary rotations (U_k) we measure the two coefficients $\langle I/Q_k \rangle \equiv \text{Tr}[\rho U_k \hat{M}_{I/Q} U_k^\dagger]$ by integrating the transmitted in-phase (I) or quadrature

5.9 3-level tomography

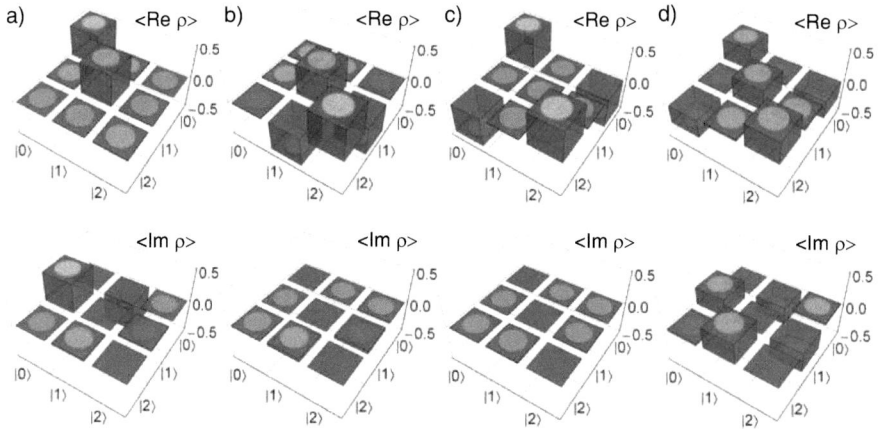

Figure 5.12 – Measured Real and imaginary part of the reconstructed density matrices of $|\Psi_a\rangle = 1/\sqrt{2}(|0\rangle + i|1\rangle)$, $|\Psi_b\rangle = 1/\sqrt{2}(|1\rangle - |2\rangle)$, $|\Psi_c\rangle = 1/\sqrt{2}(|0\rangle - |2\rangle)$ and $|\Psi_d\rangle = 1/\sqrt{3}(|0\rangle + i|1\rangle - |2\rangle)$. The cyan cylinders indicate the standard deviations, typically 0.02.

(Q) component in Eq. (5.3) over the measurement time, i.e. implementing the measurement operator $\hat{M}_{I/Q} = \int_0^T \hat{M}_{I/Q}(t)\,dt$. This relation is inverted to reconstruct the density matrix ρ by inserting the known operators $U_k \hat{M}_{I/Q} U_k^\dagger$. Note, that unlike in the preceding measurement of the populations only, we now extract a single quantity, $\langle I/Q_k \rangle$, for each measured time trace. Quantum state tomography based on the simultaneous extraction of the populations of $|0\rangle$, $|1\rangle$ and $|2\rangle$ could potentially reduce the number of required measurements, but might come at the expense of larger statistical errors, as discussed in the previous sections. The set of tomographic measurements chosen by the rotations stated in Eq. (5.11) is complete if the matrix A, defined as $\langle I/Q_k \rangle = \sum_{l=0}^{8} A_{kl} r_l$, where r_l are the coefficients of the density matrix ρ is nonsingular.

Examples of measured density matrices are shown in Fig. 5.12 for a set of four significant states. Since the tomography routine does not in general return a physical hermitian, positive definite density matrix with trace one, a maximum likelihood estimation procedure has been implemented following [James01]. This method numerically finds the density matrix that is most likely to produce the measured data assuming Gaussian noise, see Appendix B.2. The cyan cylinders indicate the standard deviation in the density matrix entries and are obtained by propagating the measured statistical standard deviations in the measurement outcomes $\hat{M}_{I/Q}$

CHAPTER 5. GENERALIZATION TO 3-LEVELS

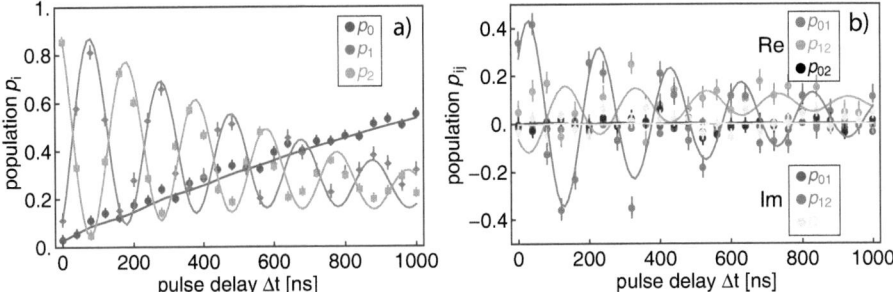

Figure 5.13 – Ramsey experiment between the $|1\rangle$ and $|2\rangle$ state. a) Diagonal entries of the density matrix. b) Real and imaginary part of the off diagonal entries of the density matrix.

to the reconstructed density matrix entries.

The extracted fidelities $F \equiv \langle \psi | \rho | \psi \rangle$ of $95 \pm 2\%$, $97 \pm 2\%$, $97 \pm 2\%$ and $92 \pm 2\%$ respectively, demonstrate the high level of control and the good understanding of the read-out of our three level system. Considering the measured decay rates, the best achievable fidelity for the states $|\Psi\rangle$ is $97 \pm 1\%$. Preparing a set of the 12 different states (comprising the basis states $|i\rangle$ and the superposition states $(|i\rangle + |j\rangle)/\sqrt{2}$, $(|i\rangle + i|j\rangle)/\sqrt{2}$, $(|0\rangle + |1\rangle + |2\rangle)/\sqrt{3}$, $(|0\rangle + |1\rangle + i|2\rangle)/\sqrt{3}$ and $(|0\rangle + i|1\rangle + i|2\rangle)/\sqrt{3}$, where $i, j = 0, 1, 2$) we measure an average fidelity of 95%, with a minimum of $92 \pm 2\%$ for the pure $|2\rangle$ state. The small remaining imperfections are likely due to phase errors in the DRAG pulses which affect both state preparation and tomography or could be due to a slightly different measurement operator from the used $\hat{M}_{I/Q}$.

The described tomographic reconstruction procedure can be used to asses any operation on the qutrit. As an example a Ramsey experiment is performed as in Sec. 5.8, driving both transitions on the y-axis. In contrast to the measurements demonstrated in Sec. 5.8, the state tomography procedure has been applied to emphasize any unwanted phase accumulated during the procedure and rule out systematic errors from the three level population reconstruction method. In this experiment the ω_1 transition is driven 5 MHz below the resonance and the energy decoherence times $T_1^{01} = 840$ ns and $T_1^{12} = 860$ ns have been extracted from a separate experiment similar the ones described in Sec. 4.6 and 5.7. The measured data, shown in Fig. 5.13, displays the same features discussed in Sec. 5.8 and is again fitted to three level Bloch equations finding $T_\phi^{12} = 1800 \pm 50$ ns. The data shown in Fig. 5.13b displays the evolution of the off diagonal entries of the density matrix during the Ramsey experiment. The data is

5.10 Outlook

in good agreement with the model, where the only non-zero matrix element $p_{ij} = \langle \Psi | i \rangle \langle j | \Psi \rangle$ is p_{12}, depicted in green (real part) and magenta (imaginary part). The Ramsey sequence should not generate any phase component in the subspace of the $\langle 0 |$, $\langle 1 |$ and $\langle 0 |$, $\langle 2 |$ states and decoherence does not generate any additional phase coherence, so the lack of any significant population in the matrix elements p_{01} and p_{02} is a further confirmation of the quality of the phase compensation implemented with the DRAG pulses.

5.10 Outlook

Populating and reading out higher excited transmon states is feasible with the methods described above. The suppressed charge dispersion of transmons, which allows for long coherence times is, however, exponentially enhanced for higher levels, see Eq. (2.29). For example, the sample used to perform the experiments described above, has a charge dispersion of the ω_0 transition of 10 kHz, a manageable charge dispersion of the ω_1 transition of 275 kHz, a problematic charge dispersion of the ω_2 transition of 5 MHz and a massive charge dispersion of the ω_3 transition of 60 MHz. 1/f noise would make the accurate addressing of the fourth level impossible and its coherence would be too short to be of any use in future quantum computation efforts.

On the other way, extending the three level read-out procedure to more qubits, in the spirit of [Filipp09], to demonstrate entanglement in a higher dimensional Hilbert space should be manageable without too much effort. It could be utilized, for example, to benchmark the gates proposed and realized in [Lanyon09]. Full quantum state tomography for larger systems is, however, impractical because of the exponential growth of the necessary measurements. To reconstruct a single two qubit density matrix up to three levels per qubit would imply 80 measurements which are still practical but a four qubit density matrix with three levels has already 6560 degrees of freedom.

The presented read-out methods, based on the linear dispersive approximation are limited to small photon numbers in the resonator, implying a low signal to noise ratio in single shot realizations using commercial HEMT amplifiers instead of quantum limited ones. A generalization to higher measurement powers could open the way to high fidelity single shot read-out schemes and are reviewed in Chapter 6.

Chapter 6

Nonlinear Cavity Response

The measurement procedure described in Chapter 4, based on the dispersive qubit read-out cannot reach high single shot read-out fidelities for common system parameters. Recent work exploiting bifurcation in a nonlinear oscillator to amplify quantum signals [Vijay09] and the single shot read-out of a transmon qubit [Mallet09], however, demonstrated the feasibility of high fidelity measurements. The key element of these read-out schemes is the bistable dynamics of a nonlinear element, where the system evolves to different macroscopic states for different initial qubit states.

The goal of this chapter is to exploit the qubit induced nonlinear cavity response as an embedded read-out device, as recently demonstrated in [Reed10a] and modeled in [Boissonneault10, Bishop10]. The cavity is strongly driven, enabling for a single shot qubit read-out with 84% fidelity. To achieve this goal, we study the nonlinear response beyond the resonant limit investigated in [Bishop09a]. The measured cavity responses near the critical photon number $n_{crit} = |\Delta_{ar}|/4g^2$ are fitted to a Duffing model, and the inferred nonlinearity is compared to an extended dispersive Hamiltonian. Finally, the single shot fidelity of high power read-out is assessed.

6.1 High power cavity response

The simplest extension to the harmonic oscillator is to consider a fourth order contribution in the potential. This leads to a cubic term in the equations of motion for the dimensionless field

6.1 High power cavity response

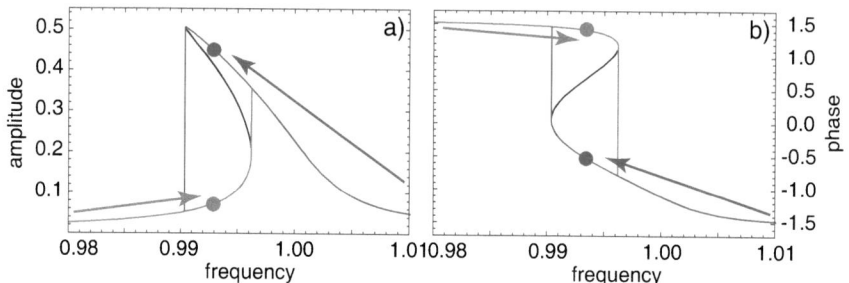

Figure 6.1 – a) Amplitude and b) phase of a Duffing oscillator. The high (blue) and low (red) amplitude solutions to Eq. (6.1) are stable. Which branch is realized depends on the drive history and is depicted by the arrows.

amplitude a [Duffing18, Landau76]

$$\ddot{a} + \omega_0^2 a = -2\mu\dot{a} - \alpha a^3 + k\cos(\Omega t), \qquad (6.1)$$

where ω_0 is the resonance frequency of the harmonic oscillator, μ the dissipation rate, α the strength of the cubic nonlinearity, k the drive amplitude and Ω the drive frequency. This equation has been extensively studied in the framework of nonlinear dynamics and chaos and is named after G. Duffing, see [Holmes76] for a review. A simple stationary solution, in the limit of small excitations is $a \approx u\cos(\Omega t - \gamma)$, has a constant amplitude u and phase γ. For a given drive frequency Ω, there is no unique correspondence between driving amplitude k and oscillation amplitude u. It is, however, possible to calculate analytically the inverted relation, finding the oscillation frequency response to a given oscillation amplitude

$$\Omega - \omega_0 = \frac{3}{8}\frac{\alpha}{\omega_0}u^2 \pm \sqrt{\frac{k^2}{4\omega_0^2 u^2} - \mu^2}, \qquad (6.2)$$

and the corresponding oscillation phase

$$\gamma = \arctan\left[\mp\frac{1}{\mu}\sqrt{\frac{k^2}{4\omega_0^2 u^2} - \mu^2}\right], \qquad (6.3)$$

which has a solution for $k^2/(4\omega_0^2 u^2) > \mu^2$. A set of solutions in the bifurcating regime is

CHAPTER 6. NONLINEAR CAVITY RESPONSE

plotted in Fig. 6.1a for the amplitude and b for the phase. The solutions can be partitioned in three branches: the lower stable branch colored in red, the higher stable branch in blue and the unstable branch with an intermediate amplitude in black. An intuitive way to realize the two distinguishable solutions is to drive the system with a constant amplitude and then change the frequency adiabatically, avoiding any sudden jumps. So if one starts with a blue detuning, the state with bigger amplitude will be realized as sketched by the blue arrow and the opposite will happen for red detuning. The same idea applies for a fixed measurement detuning and a qubit dependent resonance frequency which would be detected by a different oscillation phase and amplitude for a different initial qubit condition which would be detectable even when the qubit is decayed [Vijay09].

To experimentally asses the realizability of such a read-out procedure, we measure the resonator response in the dispersive regime with different drive powers and extract all relevant parameters. Selected traces, at fixed measurement powers are shown in Fig. 6.2a, while the full dataset is plotted in panel b). The response is nearly harmonic at low driving powers and starts to show the characteristic Duffing shape around a driving power populating the cavity with n_{crit}. The amplitude, frequency and quality factor of the resonator are extracted from a Lorentzian fit at low powers (below the first red arrow). In a second step, the entire dataset, taken at different powers was fitted to Eq. (6.2) with $\alpha = -0.1 \; 10^{-3}$ GHz^2/ Photon as the only free parameter. One can clearly observe the crossover from the harmonic response to the nonlinear case, in good agreement with the theoretical predictions. The measured response far in the nonlinear regime shows saturation and the appearance of multiple peaks, see Fig. 6.2b. For this reason data measured more than 3 dB above the power showing the first bistable response, marked with the upper red arrow, is not taken into account for the fit. In this region, further nonlinear terms must be taken into account to provide an accurate model. In the strong driving limit, above n_{crit}, a non-perturbative approach, described in [Boissonneault10] is more suited to describe the observed behavior. The data measured in the intermediate regime is, however, in good agreement with the simple Duffing model and returns useful information about the nonlinearity of the underlying system. Similar investigations, in the contest of parametric amplification where performed before [Castellanos08, Bergeal10]. In the next section a theoretical framework is developed to analyze the system behavior for arbitrary parameters.

6.2 Nonlinearity versus detuning

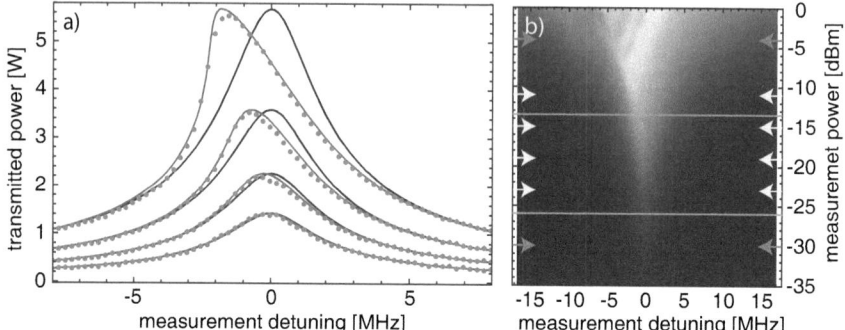

Figure 6.2 – a) Measured power transmitted through the cavity versus excitation frequency, taken at different driving powers (about 2, 5, 13 and 32 photons on average in the cavity, indicated by yellow arrows in b for a qubit detunings $\Delta/2\pi = 906$ MHz on sample 4 (see table 6.1). The black traces depict the linear Lorentzian response, while the blue curves result from a fit to the Duffing equation (6.2). The full dataset is shown in panel b, where -26 dBm correspond to 1 photon on average in the cavity on resonance (orange line) and n_{crit} is indicated by a green line. The red lines depict the region taken into account to perform the global fit.

6.2 Nonlinearity versus detuning

The simple classical picture considered in Sec. 6.1 cannot account for the source of the nonlinearity which is fundamentally quantum mechanical and arises from the transmon anharmonic level structure. The quantum anharmonic oscillator driven with a coherent field has a rich history, and has been analyzed among others by [Bhaumik75, Rozanov81, Krivoshlykov82, Bose87] and more recently by [Dykman05, Marthaler06, Dykman07, Serban07, Serban10]. So one has to reconcile the classical description, used to extract the nonlinearity, with the system Hamiltonian derived in section 2.4, which quantitatively predicts the nonlinearity.

To compare the measured α, fitted to Eq. (6.1) with the expected self Kerr coefficients K_i of Eq. (2.33), one can start from the standard Hamiltonian of a driven anharmonic oscillator

$$H = \frac{1}{2m}\hat{p}^2 + \frac{m\omega_0^2}{2}\hat{x}^2 + \frac{1}{4}\gamma\hat{x}^4 - \hat{x}A\cos\Omega t, \quad (6.4)$$

where γ is the nonlinearity constant. As a first step, we replace the generalized moment $\hat{p}$ with the charge operator $\hat{q}$, the position $\hat{x}$ with the flux operator $\hat{\phi}$ and the mass with the capacitance

CHAPTER 6. NONLINEAR CAVITY RESPONSE

C to switch to an electromagnetic picture and using $\Omega = 1/\sqrt{LC}$ and $Z_0 = \sqrt{L/C}$. The new Hamiltonian can be expressed in terms of rising and lowering operators of the oscillator, by replacing $\hat{\phi} = \sqrt{\hbar Z_0/2}(\hat{a}^\dagger + \hat{a})$, and $\hat{q} = -i\sqrt{\hbar/2Z_0}(\hat{a}^\dagger - \hat{a})$. Transforming to the rotating frame with $U(t) = \exp\left[i\Omega \hat{a}^\dagger \hat{a} t\right]$, yields the Hamiltonian

$$H_U = -\hbar(\Omega - \omega_0)\hat{n} + \frac{1}{2}\hbar V \hat{n}(\hat{n}+1) - \hbar f(\hat{a} + \hat{a}^\dagger), \quad (6.5)$$

where $V = (3\hbar\gamma)/(4\omega_0^2)$. This Hamiltonian has the same form as Eq. (2.33) and can be used to express γ in terms of K_i, finding $V/2 = \sum_i K_i |i\rangle\langle i|$. To write α in terms of γ, we calculate the equations of motion from Eq. (6.4), using $\dot{q} = -(\partial/\partial\phi)H$ and $\dot{\phi} = (\partial/\partial q)H$ and find

$$\ddot{a} + \omega_0^2 a = -2Z^2 \hbar \omega_0 \gamma a^3 + \frac{A\omega_0 \sqrt{Z}}{\sqrt{2\hbar}} \cos(\Omega t). \quad (6.6)$$

From this we find $\alpha = 2\hbar Z^2 \omega_0 \gamma$ and by comparison with Eq. (6.1) in the same units we finally obtain

$$\alpha = \frac{4}{3}\omega_0 \sum_{i=0}^{M-1} K_i |i\rangle\langle i|. \quad (6.7)$$

Eq. (6.6) is written in units of number of photons ($a^*a = n$), while the amplitudes of Eq. (6.1) are expressed in Volts. It is therefore necessary to know the average photon number n inside the cavity at a given driving power to calculate the nonlinearity. We determined n from two independent measurements: an AC-Stark calibration and an estimate of the applied microwave power at the input of the resonator, which are in good agreement.

Figure 6.3 shows the fitted nonlinearity α on a logarithmic scale in dependence of the qubit transition frequency ω_0 for four different samples whose parameters are summarized in table 6.1.

For $n \ll n_{crit}$, the linear dispersive approximation holds and no nonlinear behavior can be observed. Increasing n deteriorates the dispersive approximation of Eq. (2.23) and more terms have to be taken into account while for $n \approx n_{crit}$ the approximation starts to break down completely and excited levels are significantly populated by the photons in the resonator [Boissonneault08, Boissonneault09]. In the intermediate regime, used to fit the nonlinearity, K_1 dominates over the other nonlinear terms and is therefore used to reproduce the data.

The fitted nonlinearity versus qubit detuning, displayed in Fig. 6.3, shows a monotonically falling nonlinearity over almost three orders of magnitude with increasing detuning. In this

6.2 Nonlinearity versus detuning

	$\omega_r/2\pi$ [GHz]	$\kappa/2\pi$ [MHz]	$g_0/2\pi$ [MHz]	$E_c/2\pi$ [MHz]
Sample 1×	6.425	1.57	133	232
Sample 2∘	6.439	3.98	54	475
Sample 3◇	7.019	4.05	119	314
	6.936	3.45	111	285

Table 6.1 – Characteristic frequencies of 4 different samples used to asses the transmon induced nonlinear cavity response.

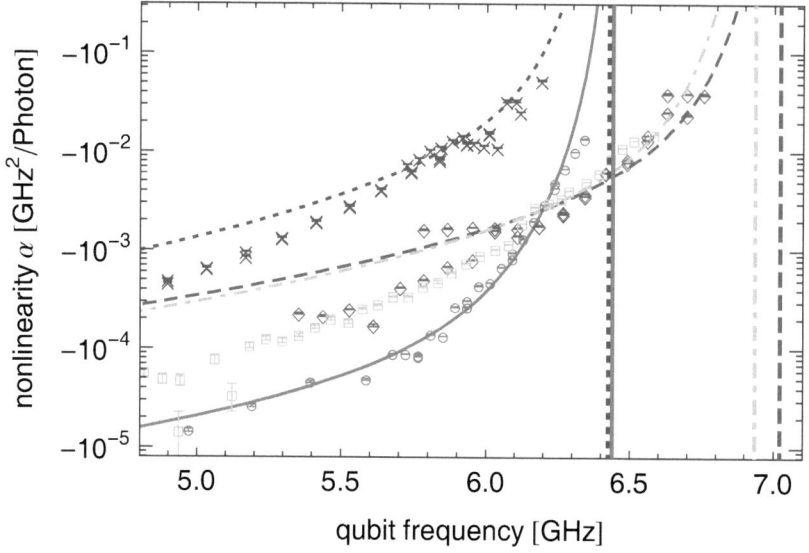

Figure 6.3 – Nonlinearity versus detuning for four different samples (see Tab. 6.1, qubit 1 - magenta crosses, qubit 2 - red dots, qubit 3 - blue diamonds, qubit 4 - cyan squares). The vertical lines indicate the frequency of the respective resonator while the traces show the expected α, calculated evaluating the single self Kerr coefficient K_1.

CHAPTER 6. NONLINEAR CAVITY RESPONSE

logarithmic plot, the sign of the nonlinearity is lost, but all Duffing resonators bend to lower frequencies, indicating a negative nonlinearity induced by qubits operated below the cavity resonance. If the qubit is operated above the resonator, however, the sign of the nonlinearity changes and the simple monotonic behavior is lost, since the higher excited states cross the resonance.

To calculate the corresponding self-Kerr coefficients K_i, used to predict the nonlinearities to lowest order, g_i is approximated by $g_i \simeq \sqrt{i+1} g_0$ and $\omega_i \simeq \omega_0 - iE_c/\hbar$. The model contains only a single adjustable parameter, namely the uncertainty of ~ 3 dB about the driving power leading to $n = 1$, corresponding to a small vertical offset in the logarithmic plot. Although it considers a single component of the self Kerr shifts K_1, the agreement between data and theory, shown in Fig. 6.3 is remarkable. To explain the remaining differences, more self-Kerr coefficients K_i must be taken into account, scaled with the respective transmon states populations induced by the photons present in the resonator. For small detunings this model breaks down and one has to solve the full master equation of the Rabi vacuum mode splitted regime and the responses must be analyzed using the same methods used in [Bishop09a]. At large detunings the nonlinearity is too small to be observed and at very high powers the perturbative model breaks down and one has to address the problem analytically [Boissonneault10].

The impossibility to drive the Duffing resonator hard enough to reach the bistable region before tipping over into an other regime makes the direct implementation of the ideas discussed in [Vijay09], where the different quantum states are distinguished via different macroscopic oscillation amplitudes/phases impractical. Driving the resonator at even higher powers, in the region where the response behaves classically allows, however, for a read-out with improved signal to noise ratio and therefore high single shot fidelity at the expense of the QND nature of the read-out.

6.3 High power qubit read-out

Measuring the resonator response at even higher powers manifests a new property of the system. If the cavity is driven in a regime corresponding to around 1'000 photons, the low power peak disappears and a new bright Lorentzian line arises at the bare cavity resonance frequency, see Fig. 6.4, with data taken in July 2009. The emergence of the new peak can be explained intuitively with a quantum to classical transition, driven by a strong coherent tone, in the spirit of [Fink10] or modeled quantitatively, as recently performed in [Boissonneault10]. Similar

6.3 High power qubit read-out

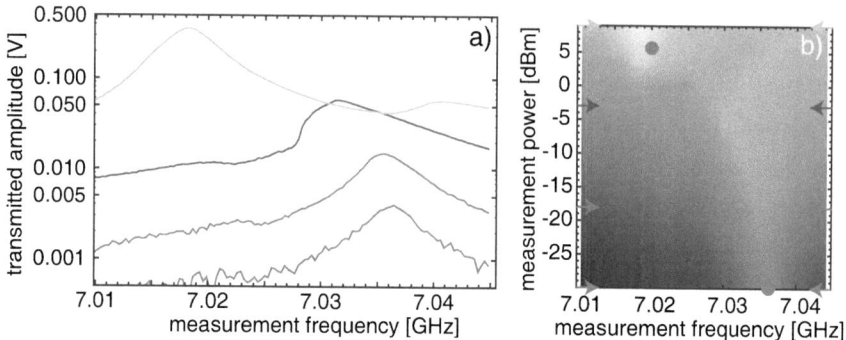

Figure 6.4 – a) Steady state cavity response for different measurement powers on a logarithmic scala, measured on sample 3 (Tab. 6.1). The selected drive powers correspond to 0.4 (red), 6 (magenta) 200 (blue) and 1'600 (cyan) photons on average in a strictly linear cavity and are indicated with colored arrows in panel b. b) Transmitted amplitude versus power and frequency. The resonator behaves linearly at low powers, goes over to the bistable intermediate Duffing regime and shows a different shifted resonance at very high drive amplitudes. The red dots indicate the power and frequency of the measurement tone used for the qubit read-out data shown in Fig. 6.5.

results as the ones described in this and in the next section have been demonstrated recently in [Reed10a].

From the measured, dispersively shifted (see Eq. (2.23)) low power resonance frequency $\omega_m/2\pi = 7.036$ GHz, fitted to Fig. 6.4, one can extract the bare cavity frequency by subtracting S_0, as demonstrated in Fig. 5.1c, described by Eq. (2.34). The transmon from sample 3 of Tab. 6.1 is operated at $\omega_0/2\pi = 6.320$ GHz, which implies $S_0 = 19$ MHz and $\omega_r/2\pi = 7.017$ GHz. So it is legitimate to interpret the measured high power resonance at $\omega_m/2\pi = 7.018$ GHz, which also has the same linewidth, as the bare cavity response. This regime can be called "classical", because in contrast to the low power limit which has to be described by the Jaynes-Cummings Hamiltonian, it is driven by a strong coherent field which can be described classically with the transmon qubit as a small perturbation. The Jaynes-Cummings nonlinearity, induced by the strong qubit-resonator coupling, which blocks the transmission at ω_r is suppressed by the classical field present in the resonator and the transmon, described as a single multilevel system, is saturated [Reed10a, Boissonneault10].

If, however, a different transmon state would delay the emergence of the classical field,

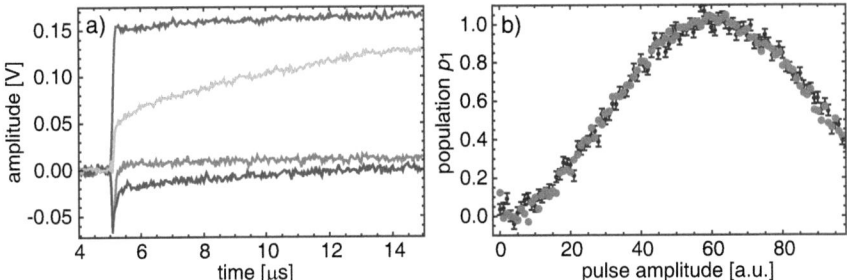

Figure 6.5 – a) High power, time resolved, pulsed response to prepared $|g\rangle$ (red, blue) and $|e\rangle$ (purple, cyan) states for the Q and I quadratures respectively, measured with +6 dBm (1'600 photons) at $\omega_m/2\pi = 7.020$ GHz. b) Comparison of low and high power read-out. Reconstructed populations for a Rabi oscillations experiment measured with -30 dBm (0.4 photons, averaged 1.3 million times) at $\omega_m/2\pi = 7.036$ GHz (black dots). The red dots are reconstructed with the area method at high power, but normalized to the measured excited state traces shown in a) instead of theoretically generated ones.

one could detect the difference in this macroscopic field and employ it as a single shot read-out device, as performed in [Castellanos08, Bergeal10]. A time-resolved measurement, performed at the same power as indicated in Fig. 6.4b with the red dot at higher power, repeated 1'000 times for a prepared ground and excited state is shown in Fig. 6.5a. In this case the experiment is not repeated many times to improve the signal to noise ratio, as discussed in Ch. 4 but to acquire enough statistics to perform an ensemble average and determine p_1. The traces are clearly distinguishable, even on a timescale which is much longer than the qubit lifetime $T_1^{01} \approx 600$ ns. The measurements go beyond the scope of the cavity-Bloch equations or the linear harmonic response and can, therefore, not be fitted to them. To compare different traces taken at different measurement detunings, the $I - Q$ traces are rotated in the same way described in Sec. 4.2, ensuring $Q = 0$ in the steady state.

Such a measurement could be used to reconstruct the qubit population using the same procedures described in Sec. 4.4, where the area between the measured curve and the ground state response is directly proportional to the population of the unknown state. The lack of a theoretical prediction for the ground and excited state responses $s^i_{g/e}(t)$ is not impeding the use of the area method described by Eq. (4.12). This is demonstrated in Fig. 6.5b, where a Rabi experiment (see Sec. 4.5) is performed and the populations are reconstructed using the well established low power read-out to generate the black dots. The same states are read-out at high power, with the parameters described above, and the populations are reconstructed using the experimentally measured $s_{g/e}^I$ (shown in Fig. 6.5a), resulting in the red points. The high power populations agree very well with the low power measurements, demonstrating the validity of the read-out and were averaged more than 1'000 times less than typical measurements shown for example in chapter 4. Acquisition time can therefore be shortened considerably.

This read-out method should not be projective in the qubit basis $|g/e\rangle$ because even much lower measurement powers induce significant populations of higher excited states of the transmon [Boissonneault09]. After a short measurement time, the transmon is excited to a superposition of higher excited states [Boissonneault10, Wilson10]. This is not a problem to read-out the system state but will destroy the state of any other qubits in the same cavity and impede any feedback protocol.

6.4 Single shot read-out

The high SNR of the data shown in Fig. 6.5a suggests the possibility of performing single shot read-out of the qubit state. While high single shot fidelities in flux qubits are routinely achieved [Chiorescu03, Cooper04, McDermott05, Lupascu06, Katz06, Lucero08] and are in reach for charge type qubits using an additional read-out device [Siddiqi06, Mallet09, Bergeal10], direct single shot read-out of a transmon using the dispersive interaction has been demonstrated only recently [Reed10a].

As discussed in Ch. 4, the outcome of an ideal single measurement of the state $|\Psi\rangle$ is "g" with probability $|c_g|^2$ and "e" with probability $|c_e|^2$. From all the information acquired during a single measurement one has therefore to define a 'decision' function which returns "g" for a prepared $|g\rangle$ state, "e" for a prepared $|e\rangle$ state and the right distributions for superposition states [Gambetta07]. A simple way to do so, is to consider a single quadrature channel, to sum the measured voltages, similarly to Eq. (4.12), up to a given time t_{end} defining a score. This has to be compared to a fixed threshold. If the summed score is below the threshold, "e" is returned else the result is "g". The scores for 10'000 single shot measurements of state $|g\rangle$ and $|e\rangle$ are shown in Fig. 6.6 in blue and red respectively.

The fidelity of such a measurement can be defined as the probability of the wrong answer

CHAPTER 6. NONLINEAR CAVITY RESPONSE

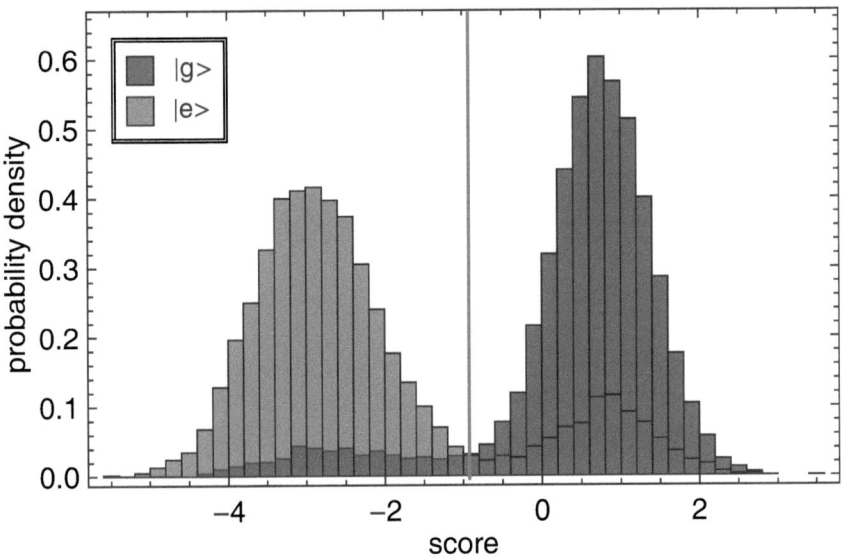

Figure 6.6 – Histogram of the measured scores for 10'000 prepared $|g\rangle$ (blue) and $|e\rangle$ (red) states. The vertical axis indicates the fraction of the measured values divided by the bin width. The pink threshold returns a fidelity of $F = 84\%$.

of the decision function [Gambetta07]:

$$F = 1 - P(\text{"}g\text{"}||e\rangle) + P(\text{"}e\text{"}||g\rangle) \approx 1 - \frac{a_{\text{"}g\text{"}}}{n_g} - \frac{a_{\text{"}e\text{"}}}{n_e}, \tag{6.8}$$

where $P(\text{"}i\text{"}||j\rangle)$ is the conditional probability of the answer "i", given the state $|j\rangle$ has been prepared, $a_{\text{"}i\text{"}}$ is the number of the wrong answers "i" and n is the number of repetitions for each prepared state.

Equation (6.8) is maximized experimentally varying different parameters, such as the measurement frequency and power, the rotation angle in the IQ-plane (a 2-dimensional approach, considering both quadratures does not significantly improve the fidelity), the integration time t_{end} and the threshold value. This procedure was performed in collaboration with C. Lang and is described in more details in [Lang09]. The data shown in Fig. 6.6 are optimized in all these

103

6.4 Single shot read-out

parameters and have a fidelity of $F = 84\%$.

The magenta states below the threshold score of -1 are prepared ground states which are misinterpreted as excited states, while the states leading to the magenta bars above the score -1 are prepared as excited states but detected as ground states. Many prepared $|e\rangle$ states decay before the detection (spontaneously or measurement induced [Boissonneault09]) and thus generate the small magenta peak of states around score 1 which is erroneously identified as "g". The strong measurement tone has also a finite probability of exciting the qubit in the first part of the measurement, generating the long negative score tail of the prepared ground states. These two error mechanisms account for most of the misidentification of the states and dominate over the statistical overlap of the distribution tails. They are, however, not relevant when reading out the state in an ensemble average, as performed in Sec. 6.3 because in that case a different average amplitude is sufficient to distinguish two states, in theory with arbitrary precision by averaging more.

As discussed in Sec. 6.3, this measurement is not projective. This is not a problem for the read-out, it is, however, a limiting factor for quantum feedback protocols such as error correction algorithms [Knill00] or measurement based state preparation [Bishop09b] because the QND property of the measurement is lost and the final state is not restricted to the same 2-dimensional Hilbert space anymore.

Appendix A

Microwave Devices at Cryogenic Temperatures

A.1 Cryogenic heat flows

For small temperature gradients, the rate of heat flow per unit area $\dot{q}$ resulting from a temperature difference ΔT in a material of cross-section A is given by [Pobell06]

$$\dot{q} = \dot{Q}/A = -\kappa \cdot \Delta T, \tag{A.1}$$

where the thermal conductivity κ is assumed to be constant. It can be generalized to

$$\dot{Q} = \frac{A}{L}\int_0^L \dot{q}dx = \frac{A}{L}\int_{T_1}^{T_2} \kappa(T)dT := -\frac{A}{L}\Theta_{T_1}^{T_2} \tag{A.2}$$

for a solid of length L with temperatures T_1 and T_2 at its ends. $\Theta_{T_1}^{T_2}$ is called thermal conductivity integral and can be calculated analytically in certain limits or inferred from specific tables [Pobell06, Lake-Shore04]. For an insulating material at less then one tenth of the Debye temperature T_D, the main contribution to the thermal conductivity is of phononic nature, with $\kappa_{phonon} = bT^3$, implying $\Theta_{T_1}^{T_2} = -b/4(T_2^4 - T_1^4)$. For metals below around 10 K, electrons dominate the thermal conductivity, implying $\kappa_{el} = \kappa_0 T$ and $\Theta_{T_1}^{T_2} = -\kappa_0/2(T_2^2 - T_1^2)$. The most relevant heat conductivities, needed to wire a dilution refrigerator are tabulated in Tab. A.1.

To wire a dilution refrigerator without affecting its base temperature, one must ensure that the combined head load on each plate from all the added wires does not exceed the cooling power on that stage. The typical cooling power available on different temperature

A.1 Cryogenic heat flows

Constant	Copper	Stainless steel	Teflon
Θ_4^{300} [W/m]	$1.5 \cdot 10^5$	2'500	100
Θ_{60}^{300} [W/m]	$5 \cdot 10^4$	2'000	100
Θ_4^{60} [W/m]	$1 \cdot 10^5$	200	10
κ_0 [W m^{-1} K^{-1}]	100	0.15	
b [W m^{-1} K^{-1}]			$5 \cdot 10^{-4}$

Table A.1 – Thermal conductivity of solids frequently used in low temperature applications

Temperature	Vericold DR-200	Oxford 400HA
70 K	4 W	-
4 K	0.5 W	>1 W
1.6 K	-	>10 mW
Still	1 mW	1 mW
100 mK	~20 μW	~20 μW
base	<1 μW	<1 μW

Table A.2 – Cooling powers on different temperature stages

stages of the cryostats are listed in Tab. A.2. UT-85 semirigid coaxial cables have a surface area of $3.6 \cdot 10^{-6}$ and $2.1 \cdot 10^{-7}$ m^2 for the shield and center conductor respectively which combined with the actual cable length and material determine the heat-flow. They usually use teflon as insulating material which at low temperature has a very bad conductivity, see Tab. A.1 and can be neglected while calculating heat-flows. However, Teflon contracts more then the surrounding metals when cooled down, possibly making a bad thermal contact with either the outer or inner conductor. Therefore, to ensure a complete thermalization of the center-conductor of a coaxial cable an attenuator (which connects the center-conductor to the ground with a resistor) or a circulator must be employed. Attenuators are usually thermalized with a copper clamp connected to the cryostat via an unfluxed copper-braid, while bulkier components such as amplifiers or circulators have their own copper mounts. If no such measure is implemented only the outer conductor can be thermlized reliably, while the inner conductor must be assumed to thermalize only at the next stage and therefore thermally load a lower cryostat plate.

Furthermore, to avoid big strains on the connectors due to the thermal expansion of the cables, they are bended such that they can move, absorbing the tension.

CHAPTER A. MICROWAVE DEVICES AT CRYOGENIC TEMPERATURES

A.2 One-dimensional black body radiation

Coaxial cables and transmission lines provide a single degree of freedom for a propagating electromagnetic field, so they can be approximated as being one-dimensional. The spectral energy density emitted at frequency ν by a black body at temperature T is given by

$$S = \frac{2h\nu}{e^{\frac{h\nu}{k_B T}} - 1}, \quad (A.3)$$

where k_B is the Boltzmann constant. Similarly for an artificial atom with $n+1$ energy levels, the average thermal population P_i of a given energy level E_i in equilibrium is given by

$$P_i = \frac{e^{-\frac{E_i}{k_B T}}}{\sum_{j=0}^{n} e^{-\frac{E_j}{k_B T}}}, \quad (A.4)$$

where $E_0 = 0$.

To have an idea of the energy scales, 50 mK correspond to 1 GHz, so the average thermal population of the first excited state with transition frequency $\omega_0/2\pi = 5$ GHz in thermal equilibrium at 50 and 30 mK is 1 and 0.03% respectively. Therefore, to manipulate a qubit with a precision higher then 1%, it must be in thermal equilibrium below 50 mK and should not be exposed to any radiative field at higher temperature in the band of the relevant energy transitions.

Again at 5 GHz, the flux of photons calculated with Eq. (A.3), generated at 300 K is 80 times more intense then the one generated at 4 K. To attenuate this noise, a 20 dB attenuator (absorbing a factor of 100 in power) is mounted on the 4 K stage, see Fig. 3.2. The radiation emitted at 100 mK is in turn 160 times weaker then the power emitted at the 4 K plate, justifying the presence of a second 20 dB attenuator. The remaining radiation corresponds to around 0.1 photons in average in the resonator which are further suppressed with a last 10 dB attenuator placed at base temperature.

A.3 Circulators

Circulators are used as isolators, routing the weak measurement signal from the sample to the cold amplifier, while absorbing the incoming signal in the third port with a matched termi-

A.3 Circulators

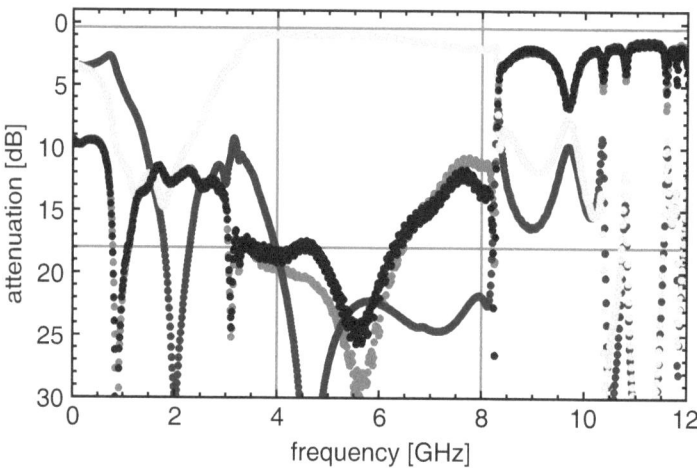

Figure A.1 – Measured S-parameters of the Pamtech circulator #110. S_{11} S_{12} S_{21} S_{22} are plotted in red, blue, green and black respectively. The specified working figures are shown as gray lines.

nation. They use a permanent magnet made of ferrite to break the symmetry of the system, so one should avoid placing them in vicinity of the experiment because of possible magnetic disturbance.

Pamtech (now owned by QuinStar Technology) produces cryogenic circulators working in the bands 4-8 and 6-12 GHz with 18 dB isolation and 0.4 dB insertion loss. They also offer double isolators with higher bandwidths 4-12 GHz but with reduced isolation and higher insertion loss. 4-8 GHz circulators with identical specifications are also offered by Raditek Inc.

A sample 4-8 GHz circulator from Pamtech (#110) is measured at 4 K, resulting in the scattering matrix parameters (S-parameters) plotted in Fig. A.1. Port 3 of the circulator was terminated with a matched 50 Ohm impedance. The measured forward transmission S_{21}, plotted in green displays the expected flat band between 4 and 8 GHz, while the inverse measurement of S_{12}, displayed in blue shows at least 18 dB attenuation at the lower end of the frequency window but usually performs much better. The calibration of the cold coaxial cables depends on the liquid helium level in the dewar used to perform the measurement and is accurate only up to about 1 dB. The calibration could therefore be responsible for the slightly higher observed insertion loss then the one specified. The reflected signals S_{11} and S_{22} plotted in red and black, respectively, show at most 10 dB return loss which could come from the used cable connectors and must therefore not be ascribed to the circulator.

It is worthwhile to note that outside the specified frequency window not only S_{21} can be large, impeding a measurement, but also S_{12} can be small (only 10 dB at 3 and 10 GHz). Operating a qubit or resonator around these frequencies results in increased black body radiation transmitted from room temperature to the sample.

A.4 Copper powder filters

Copper powder filters are used as low pass filters working at cryogenic temperatures [Martinis87, Fukushima97]. Their main advantage over conventional lumped element filters is the very high attenuation at high frequency without the presence of any resonance up to at least 40 GHz. A smooth low frequency cutoff is warranted by a lumped element π-shaped RLC filter while the high frequency damping is provided by the skin effect [Lukashenko08].

A twin housing for two filters is sketched in Fig. A.2. The employed constantan wire and the stainless steel powder are resistive at low temperatures, ensuring an optimal filtering over a broad range of temperatures. The chosen π geometry with a capacitor at each end of the filter implies a second order RC filter which is confirmed by the measured filtering slope at 4 Kelvin, shown in Fig. A.3. The considered set of filters were manufactured and measured by Andreas Fragner during his Diploma thesis at ETH Zurich. The filters where assembled following the recipe:

- Solder a 100 nF surface-mount (SMD) capacitor directly on the inside of an SMA press mount receptacle from Delta electronics.

- Mix three parts in mass of Alfa Aesar 325 mesh stainless steel powder (warning: toxic, wear latex gloves, breathing mask and protective clothing) into one part of Emerson and Cumming Stycast 1266.

- Pump out the mixture to eliminate enclosed air-bubbles.

- Fill a 3 mm inner diameter tube with the mixtures and let it dry for 24 hours (or speed cure at 60 degrees Celsius for 2 hours) ensuring no curvature appears in the future winding rods.

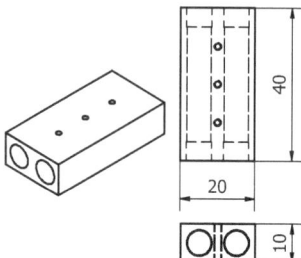

Figure A.2 – Copper housing sketch used to manufacture two copper powder filters. The dimensions are in millimeters.

- Remove the tube and cut 30 mm long rods.
- Wind 500 mm of 0.2 mm diameter constantan wire on the rod.
- Solder one SMA connector to the wire and push it into the copper case shown in Fig. A.2. (there is a press machine in the machine shop and a fitting for our connectors). Make sure the soldered, exposed end does not touch the housing, resulting in a short to ground.
- Solder the other SMA connector to the constantan wire.
- Fill the housing with evacuated SS-stycast mixture using a big syringe.
- Press in the other connector.
- Cure at 60 degrees Celsius for 2 hours in the oven located in the Printraum

A.5 Low noise power supply

The low noise HEMT amplifiers need both a stable and a carefully adjustable voltage supply. We operate the LNC4-8A amplifier from Low Noise Factory with a specified bandwidth of 4-8 GHz, a typical noise temperature of 2.6 K and 40 dB gain. They need a single gate voltage of -1 to -2 V and a source drain voltage of 0.5 to 1.5 V, depending on the actual device and operation temperature. We also operate several amplifiers from the California institute of Technology with a bandwidth of 2-12 GHz, a typical noise temperature of 4 K and 36 dB

CHAPTER A. MICROWAVE DEVICES AT CRYOGENIC TEMPERATURES

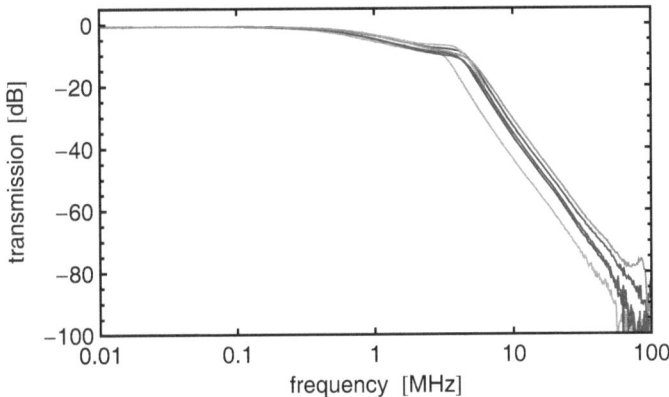

Figure A.3 – Measured transmitted power in dB trough a set of filters manufactured using the given recipe. The 3 dB point is located around 3 MHz.

gain. They need two separate gate voltages of 1.2 to 2.2 V and a source drain voltage of 1.2 to 1.8 V, depending on the actual device and operation temperature. The stated voltages must be adjusted within 10 mV and kept as constant as possible while the current flowing through the source drain must be monitored to ensure the correct operation.

The solution found and implemented by the Elektronik-Lehrlabor (ELL) at ETH design a power supply with three transformers, arranged to strongly suppress 50 Hz noise and a separate low noise circuit used to source the amplifiers. The schematics are shown in the following two pages (only online).

A.6 Switch electronics

Mechanical RF switches used at low temperature need special control electronics because the switching coils have different resistances at 4 Kelvin then at room temperature. Therefore they cannot be voltage driven, as designed by the manufacturer but need to be current driven. Furthermore, the electronics must be decoupled from the PC ground to avoid ground-loops.

To control the switch from the PC, the Low-Cost Multifunction DAQ USB-6009 from National Instrument is used in combination with a differential 30 V lab voltage source. To switch

A.6 Switch electronics

the 6 to 1 channel Radiall mechanical switch R573-423-600, 15 ms pulses carrying 125 mA are necessary (there is a reset line which switches all channels simultaneously, draining 750 mA). The room temperature resistance is about 220 Ohm, implying a specified applied voltage of 28 V. The current is regulated using a Zener diode and a reference resistor, switched with a power transistor. The grounds are decoupled using an opto-isolator. Additional resistors are added to control the voltages on the opto-isolator and a diode is put in parallel with the switch coil to drain the current when the transistor switches off. Six identical circuits control the six channels while an additional circuit is used as a global reset (with a smaller reference resistance). The final design, PCB layout and routing was drawn using the open source KiCad package. The circuit contains the following components (labeled as in the sketches):

- 7x $R_{1i} \rightarrow$ 470 Ohm
- 7x $R_{2i} \rightarrow$ 470 Ohm
- 6x $R_{3i} \rightarrow$ 12 Ohm
- 1x $R_{31} \rightarrow$ 2 Ohm
- 7x opto Darlington$\rightarrow$ 4N32
- 7x PNP$\rightarrow$ BD900A
- 7x $D_{1i} \rightarrow$ 1N4728A
- 7x $D_{2i} \rightarrow$ 1N4005

The circuit was designed, implemented and tested successfully. It cannot control only the 6 channel R573-423-600 but also the 2 channel R572.433.000. Note that the standard +5 V biased by the DAQ after each PC reboot result in a constant current flowing simultaneously in all channels of the switch. It is therefore necessary to initialize the DAQ first and then turn on the external voltage supply or an inverting circuit could be preposed to the controller.

The schematics and implemented PCB are shown in the next two pages (only online).

Appendix B

Numerical Recipes

B.1 Three level cavity-Bloch equations

The full Cavity-Bloch equations for 3 levels, considering the master equation (5.1) and Hamiltonian (5.2) with three driving terms ϵ_{01}, ϵ_{12}, ϵ_{02} at the respective frequencies $\omega_s^0 \approx \omega_0$, $\omega_s^1 \approx \omega_1$ and $\omega_s^2 = \omega_0 + \omega_1$ leading to the detunings $\Delta_s^1 = \omega_0 - \omega_s^0$ and $\Delta_s^2 = \omega_1 - \omega_s^1$ are stated below.

As in Sec. 4.1, the terms $\langle \hat{a}^\dagger \hat{a} \hat{P}_{ij} \rangle \approx \langle \hat{a}^\dagger \hat{a} \rangle \langle \hat{P}_{ij} \rangle$ and $\langle \hat{a}^\dagger \hat{a} \hat{a} \hat{P}_{ij} \rangle \approx \langle \hat{a}^\dagger \hat{a} \rangle \langle \hat{a} \hat{P}_{ij} \rangle$ are factored, while the terms $\langle \hat{a} \hat{P}_{ij} \rangle$ are kept, defining $|i\rangle\langle j| = \hat{P}_{ij}$. For a coherent field with complex amplitude α, defined by the state $|\alpha\rangle_C$ and a Fock state with n photons, defined by the state $|n\rangle_F$ this leads to the correct expressions for the factored terms

$$\begin{aligned}\langle \hat{a}^\dagger \hat{a} \hat{P}_{ij} \rangle_C &= |\alpha|^2 \langle \hat{P}_{ij} \rangle = \langle \hat{a}^\dagger \hat{a} \rangle_C \langle \hat{P}_{ij} \rangle, \\ \langle \hat{a}^\dagger \hat{a} \hat{P}_{ij} \rangle_F &= n \langle \hat{P}_{ij} \rangle = \langle \hat{a}^\dagger \hat{a} \rangle_F \langle \hat{P}_{ij} \rangle \neq \langle \hat{a}^\dagger \rangle_F \langle \hat{a} \hat{P}_{ij} \rangle_F = 0. \end{aligned} \quad \text{(B.1)}$$

B.1 Three level cavity-Bloch equations

$$d_t \langle \hat{a} \rangle = -\kappa/2 \langle \hat{a} \rangle - i \left(\epsilon_m + \Delta_{rm} \langle \hat{a} \rangle + S_0 \langle \hat{a} \hat{P}_{00} \rangle + S_1 \langle \hat{a} \hat{P}_{11} \rangle + S_2 \langle \hat{a} \hat{P}_{22} \rangle \right),$$

$$d_t \langle \hat{a}^\dagger \hat{a} \rangle = -2\epsilon_m \Im \langle \hat{a} \rangle - \kappa \langle \hat{a}^\dagger \hat{a} \rangle,$$

$$d_t \langle \hat{P}_{00} \rangle = -i \left(\epsilon_{01}^\star \langle \hat{P}_{01} \rangle - \epsilon_{01} \langle \hat{P}_{10} \rangle + \epsilon_{02}^\star \langle \hat{P}_{02} \rangle - \epsilon_{02} \langle \hat{P}_{20} \rangle \right) + \gamma_1^1 \langle \hat{P}_{11} \rangle,$$

$$d_t \langle \hat{P}_{11} \rangle = i \left(\epsilon_{01}^\star \langle \hat{P}_{01} \rangle - \epsilon_{01} \langle \hat{P}_{10} \rangle - \epsilon_{12}^\star \langle \hat{P}_{12} \rangle + \epsilon_{12} \langle \hat{P}_{21} \rangle \right) - \gamma_1^1 \langle \hat{P}_{11} \rangle$$
$$+ \gamma_1^2 \langle \hat{P}_{22} \rangle,$$

$$d_t \langle \hat{P}_{01} \rangle = -i \left\{ \epsilon_{12}^\star \langle \hat{P}_{02} \rangle + \epsilon_{01} \left(\langle \hat{P}_{00} \rangle - \langle \hat{P}_{11} \rangle \right) + \epsilon_{02} \langle \hat{P}_{21} \rangle + \langle \hat{P}_{01} \rangle \left[\Delta_s^1 \right. \right.$$
$$\left. \left. + \langle \hat{a}^\dagger \hat{a} \rangle (-S_0 + S_1) \right] \right\} - \gamma_1^1/2 \langle \hat{P}_{01} \rangle - \gamma_\phi^1 \langle \hat{P}_{01} \rangle,$$

$$d_t \langle \hat{P}_{10} \rangle = i \left\{ \epsilon_{01}^\star \left(\langle \hat{P}_{00} \rangle - \langle \hat{P}_{11} \rangle \right) - \epsilon_{02}^\star \langle \hat{P}_{12} \rangle + \epsilon_{12} \langle \hat{P}_{20} \rangle + \langle \hat{P}_{10} \rangle \left[\Delta_s^1 \right. \right.$$
$$\left. \left. + \langle \hat{a}^\dagger \hat{a} \rangle (-S_0 + S_1) \right] \right\} - \gamma_1^1/2 \langle \hat{P}_{10} \rangle - \gamma_\phi^1 \langle \hat{P}_{10} \rangle,$$

$$d_t \langle \hat{P}_{12} \rangle = i \left\{ \epsilon_{01}^\star \langle \hat{P}_{02} \rangle - \epsilon_{02} \langle \hat{P}_{10} \rangle - \epsilon_{12} \left(\langle \hat{P}_{11} \rangle - \langle \hat{P}_{22} \rangle \right) + \langle \hat{P}_{12} \rangle \left[\Delta_s^1 \right. \right.$$
$$\left. \left. - \Delta_s^2 \langle \hat{a}^\dagger \hat{a} \rangle (S_1 - S_2) \right] \right\} - \gamma_1^1/2 \langle \hat{P}_{12} \rangle - \gamma_1^2/2 \langle \hat{P}_{12} \rangle - \gamma_\phi^1 \langle \hat{P}_{12} \rangle$$
$$- \gamma_\phi^2 \langle \hat{P}_{12} \rangle,$$

$$d_t \langle \hat{P}_{21} \rangle = i \left\{ \epsilon_{02}^\star \langle \hat{P}_{01} \rangle - \epsilon_{01} \langle \hat{P}_{20} \rangle + \epsilon_{12}^\star \left(\langle \hat{P}_{11} \rangle - \langle \hat{P}_{22} \rangle \right) + \langle \hat{P}_{21} \rangle \left[-\Delta_s^1 \right. \right.$$
$$\left. \left. + \Delta_s^2 \langle \hat{a}^\dagger \hat{a} \rangle (-S_1 + S_2) \right] \right\} - \gamma_1^1/2 \langle \hat{P}_{21} \rangle - \gamma_1^2/2 \langle \hat{P}_{21} \rangle - \gamma_\phi^1 \langle \hat{P}_{21} \rangle$$
$$- \gamma_\phi^2 \langle \hat{P}_{21} \rangle,$$

$$d_t \langle \hat{P}_{22} \rangle = i \left(\epsilon_{02}^\star \langle \hat{P}_{02} \rangle + \epsilon_{12}^\star \langle \hat{P}_{12} \rangle - \epsilon_{02} \langle \hat{P}_{20} \rangle - \epsilon_{12} \langle \hat{P}_{21} \rangle \right) - \gamma_1^2 \langle \hat{P}_{22} \rangle,$$

$$d_t \langle \hat{P}_{02} \rangle = -i \left\{ \epsilon_{12} \langle \hat{P}_{01} \rangle - \epsilon_{01} \langle \hat{P}_{12} \rangle + \epsilon_{02}^\star \left(\langle \hat{P}_{00} \rangle - \langle \hat{P}_{22} \rangle \right) - \langle \hat{P}_{02} \rangle \left[\Delta_s^2 \right. \right.$$
$$\left. \left. + \langle \hat{a}^\dagger \hat{a} \rangle (S_0 - S_2) \right] \right\} - \gamma_1^2/2 \langle \hat{P}_{02} \rangle - \gamma_\phi^2 \langle \hat{P}_{02} \rangle,$$

$$d_t \langle \hat{P}_{20} \rangle = i \left\{ \epsilon_{12}^\star \langle \hat{P}_{10} \rangle - \epsilon_{01}^\star \langle \hat{P}_{21} \rangle + \epsilon_{02}^\star \left(\langle \hat{P}_{00} \rangle - \langle \hat{P}_{22} \rangle \right) - \langle \hat{P}_{20} \rangle \left[\Delta_s^2 \right. \right.$$
$$\left. \left. + \langle \hat{a}^\dagger \hat{a} \rangle (S_0 - S_2) \right] \right\} - \gamma_1^2/2 \langle \hat{P}_{20} \rangle - \gamma_\phi^2 \langle \hat{P}_{20} \rangle,$$

CHAPTER B. NUMERICAL RECIPES

$$\begin{aligned}
d_t\langle \hat{a}\hat{P}_{00}\rangle &= -\kappa/2\langle \hat{a}\hat{P}_{00}\rangle - i\left\{\epsilon_m\langle \hat{P}_{00}\rangle + \left[\epsilon^*_{01}\langle \hat{a}\hat{P}_{01}\rangle \right.\right.\\
&\quad \left.\left. +\epsilon^*_{02}\langle \hat{a}\hat{P}_{02}\rangle - \epsilon_{01}\langle \hat{a}\hat{P}_{10}\rangle - \epsilon_{02}\langle \hat{a}\hat{P}_{20}\rangle + \langle \hat{a}\hat{P}_{00}\rangle(\Delta_{rm}+S_0)\right]\right\}\\
&\quad +\gamma_1^1\langle \hat{a}\hat{P}_{11}\rangle,\\
d_t\langle \hat{a}\hat{P}_{11}\rangle &= -\kappa/2\langle \hat{a}\hat{P}_{11}\rangle - i\left\{\epsilon_m\langle \hat{P}_{11}\rangle + \left[\epsilon^*_{01}\langle \hat{a}\hat{P}_{01}\rangle - \epsilon_{01}\langle \hat{a}\hat{P}_{10}\rangle\right.\right.\\
&\quad \left.\left. -\epsilon^*_{12}\langle \hat{a}\hat{P}_{12}\rangle + \epsilon_{12}\langle \hat{a}\hat{P}_{21}\rangle + \langle \hat{a}\hat{P}_{11}\rangle(\Delta_{rm}+S_1)\right]\right\} - \gamma_1^1\langle \hat{a}\hat{P}_{11}\rangle\\
&\quad +\gamma_1^2\langle \hat{a}\hat{P}_{22}\rangle,\\
d_t\langle \hat{a}\hat{P}_{01}\rangle &= -\kappa/2\langle \hat{a}\hat{P}_{01}\rangle - i\left\{\epsilon_m\langle \hat{P}_{01}\rangle + \left[\epsilon^*_{12}\langle \hat{a}\hat{P}_{02}\rangle + \epsilon_{01}\left(\langle \hat{a}\hat{P}_{00}\rangle - \langle \hat{a}\hat{P}_{11}\rangle\right)\right.\right.\\
&\quad \left. -\epsilon_{02}\langle \hat{a}\hat{P}_{21}\rangle + \langle \hat{a}\hat{P}_{01}\rangle\left(\Delta_{rm}+\Delta_s^1+S_1\right) + \left(\langle \hat{a}^\dagger \hat{a}\rangle\langle \hat{a}\hat{P}_{01}\rangle\right.\right.\\
&\quad \left.\left.\left. +\langle \hat{a}\hat{P}_{01}\rangle\right)(-S_0+S_1)\right]\right\} - \gamma_1^1/2\langle \hat{a}\hat{P}_{01}\rangle - \gamma_\phi^1\langle \hat{a}\hat{P}_{01}\rangle,\\
d_t\langle \hat{a}\hat{P}_{10}\rangle &= -\kappa/2\langle \hat{a}\hat{P}_{10}\rangle + i\left\{-\epsilon_m\langle \hat{P}_{10}\rangle + \left[\epsilon^*_{01}\left(\langle \hat{a}\hat{P}_{00}\rangle - \langle \hat{a}\hat{P}_{11}\rangle\right)\right.\right.\\
&\quad \left. -\epsilon^*_{02}\langle \hat{a}\hat{P}_{12}\rangle + \epsilon_{12}\langle \hat{a}\hat{P}_{20}\rangle - \langle \hat{a}\hat{P}_{10}\rangle\left(\Delta_{rm}-\Delta_s^1+S_0\right) - \right.\\
&\quad \left.\left. \left(\langle \hat{a}^\dagger \hat{a}\rangle\langle \hat{a}\hat{P}_{10}\rangle + \langle \hat{a}\hat{P}_{10}\rangle\right)(S_0-S_1)\right]\right\} - \gamma_1^1/2\langle \hat{a}\hat{P}_{10}\rangle - \gamma_\phi^1\langle \hat{a}\hat{P}_{10}\rangle,\\
d_t\langle \hat{a}\hat{P}_{12}\rangle &= -\kappa/2\langle \hat{a}\hat{P}_{12}\rangle + i\left\{-\epsilon_m\langle \hat{P}_{12}\rangle + \left[\epsilon^*_{01}\langle \hat{a}\hat{P}_{02}\rangle - \epsilon_{02}\langle \hat{a}\hat{P}_{10}\rangle\right.\right.\\
&\quad \left. -\epsilon_{12}\left(\langle \hat{a}\hat{P}_{11}\rangle - \langle \hat{a}\hat{P}_{22}\rangle\right) - \langle \hat{a}\hat{P}_{12}\rangle\left(\Delta_{rm}-\Delta_s^1+\Delta_s^2+S_2\right)\right.\\
&\quad \left.\left. -\left(\langle \hat{a}^\dagger \hat{a}\rangle\langle \hat{a}\hat{P}_{12}\rangle + \langle \hat{a}\hat{P}_{12}\rangle\right)(-S_1+S_2)\right]\right\} - \gamma_1^1/2\langle \hat{a}\hat{P}_{12}\rangle\\
&\quad -\gamma_1^2/2\langle \hat{a}\hat{P}_{12}\rangle - \gamma_\phi^1\langle \hat{a}\hat{P}_{12}\rangle - \gamma_\phi^2\langle \hat{a}\hat{P}_{12}\rangle,\\
d_t\langle \hat{a}\hat{P}_{21}\rangle &= -\kappa/2\langle \hat{a}\hat{P}_{21}\rangle + i\left\{-\epsilon_m\langle \hat{P}_{21}\rangle + \left[\epsilon^*_{02}\langle \hat{a}\hat{P}_{01}\rangle - \epsilon_{01}\langle \hat{a}\hat{P}_{20}\rangle\right.\right.\\
&\quad \left. +\epsilon^*_{12}\left(\langle \hat{a}\hat{P}_{11}\rangle - \langle \hat{a}\hat{P}_{22}\rangle\right) - \langle \hat{a}\hat{P}_{21}\rangle\left(\Delta_{rm}+\Delta_s^1-\Delta_s^2+S_1\right)\right.\\
&\quad \left.\left. -\left(\langle \hat{a}^\dagger \hat{a}\rangle\langle \hat{a}\hat{P}_{21}\rangle + \langle \hat{a}\hat{P}_{21}\rangle\right)(S_1-S_2)\right]\right\} - \gamma_1^1/2\langle \hat{a}\hat{P}_{21}\rangle\\
&\quad -\gamma_1^2/2\langle \hat{a}\hat{P}_{21}\rangle - \gamma_\phi^1\langle \hat{a}\hat{P}_{21}\rangle - \gamma_\phi^2\langle \hat{a}\hat{P}_{21}\rangle,
\end{aligned}$$

B.2 Maximum likelihood estimation

$$\begin{aligned}
d_t\langle\hat{a}\hat{P}_{22}\rangle &= -\kappa/2\langle\hat{a}\hat{P}_{22}\rangle + i\left\{-\epsilon_m\langle\hat{P}_{22}\rangle + \left[-\epsilon_{02}^\star\langle\hat{a}\hat{P}_{02}\rangle - \epsilon_{12}^\star\langle\hat{a}\hat{P}_{12}\rangle + \right.\right.\\
&\quad \left.\left. \epsilon_{02}\langle\hat{a}\hat{P}_{20}\rangle + \epsilon_{12}\langle\hat{a}\hat{P}_{21}\rangle - \langle\hat{a}\hat{P}_{22}\rangle\left(\Delta_{rm}+S_2\right)\right]\right\} - \gamma_1^2\langle\hat{a}\hat{P}_{22}\rangle,\\
d_t\langle\hat{a}\hat{P}_{02}\rangle &= -\kappa/2\langle\hat{a}\hat{P}_{02}\rangle - i\left\{\epsilon_m\langle\hat{P}_{02}\rangle + \left[\epsilon_{12}\langle\hat{a}\hat{P}_{01}\rangle - \epsilon_{01}\langle\hat{a}\hat{P}_{12}\rangle + \right.\right.\\
&\quad \epsilon_{02}\left(\langle\hat{a}\hat{P}_{00}\rangle - \langle\hat{a}\hat{P}_{22}\rangle\right) + \langle\hat{a}\hat{P}_{02}\rangle\left(\Delta_{rm}+\Delta_s^2+S_2\right) + \\
&\quad \left.\left. \left(\langle\hat{a}^\dagger\hat{a}\rangle\langle\hat{a}\hat{P}_{02}\rangle + \langle\hat{a}\hat{P}_{02}\rangle\right)(-S_0+S_2)\right]\right\} - \gamma_1^2/2\langle\hat{a}\hat{P}_{02}\rangle - \gamma_\phi^2\langle\hat{a}\hat{P}_{02}\rangle,\\
d_t\langle\hat{a}\hat{P}_{20}\rangle &= -\kappa/2\langle\hat{a}\hat{P}_{20}\rangle + i\left\{-\epsilon_m\langle\hat{P}_{20}\rangle + \left[\epsilon_{12}^\star\langle\hat{a}\hat{P}_{10}\rangle - \epsilon_{01}^\star\langle\hat{a}\hat{P}_{21}\rangle\right.\right.\\
&\quad +\epsilon_{02}^\star\left(\langle\hat{a}\hat{P}_{00}\rangle - \langle\hat{a}\hat{P}_{22}\rangle\right) - \langle\hat{a}\hat{P}_{20}\rangle\left(\Delta_{rm}-\Delta_s^2+S_0\right) - \\
&\quad \left.\left. \left(\langle\hat{a}^\dagger\hat{a}\rangle\langle\hat{a}\hat{P}_{20}\rangle + \langle\hat{a}\hat{P}_{20}\rangle\right)(S_0-S_2)\right]\right\} - \gamma_1^2/2\langle\hat{a}\hat{P}_{20}\rangle - \gamma_\phi^2\langle\hat{a}\hat{P}_{20}\rangle,
\end{aligned}$$

B.2 Maximum likelihood estimation

The density matrix $\tilde{\rho}$ is reconstructed by inverting the relations

$$n_k := \langle I/Q_k\rangle \equiv \text{Tr}[\tilde{\rho}U_k\hat{M}_{I/Q}U_k^\dagger], \qquad (B.2)$$

where n_k is the measurement outcome and $U_k\hat{M}_{I/Q}U_k^\dagger$ is known ($\text{Tr}[\rho_i\hat{M}_{I/Q}(t)] = s_I^{I/Q}$ is calculated solving the cavity-Bloch equations and U_k is stated in Eq. (5.11)). The found density matrix $\tilde{\rho}$ must, however, not be a hermitian positive semidefinite matrix with trace 1 and can therefore be nonphysical.

The maximum likelihood estimation finds a physical matrix ρ such that the measured n_k are most likely to be generated by Gaussian distributed noise [James01]. A physical density matrix ρ can always be written as

$$\rho = \frac{T^\dagger T}{\text{Tr}[T^\dagger T]}, \qquad (B.3)$$

where all the t_i are real and for a three level system

$$T = \begin{pmatrix} t_1 & 0 & 0 \\ t_4 + it_5 & t_2 & 0 \\ t_8 + it_9 & t_6 + it_7 & t_3 \end{pmatrix}. \qquad (B.4)$$

The likelihood function L calculates the logarithm of the probability P of obtaining a set of n_k with a given T

$$L(t_1, t_2, ..., t_9) = \sum_i \frac{\left(\text{Tr}[\rho U_i \hat{M}_{I/Q} U_i^\dagger] - n_i\right)^2}{\sigma_i^2}, \tag{B.5}$$

where σ_i are the observed standard deviations of the measurement outcomes n_i. Maximizing the probability P is equivalent to minimizing the likelyhood function L in the variables t_i. To minimize Eq. (B.5) numerically, an initial guess for ρ is necessary. The nearest symmetric positive semidefinite matrix ρ_0 to the already known $\tilde{\rho}$ in the Frobenius norm ($\|A\|_F = (\sum_{ij} |a_{ij}|^2)^{1/2}$) is used [Higham86, Higham88]. ρ_0 can be expressed as

$$\rho_0 = \left[\left(\tilde{\rho} + \tilde{\rho}^\dagger\right)/2 + \sqrt{\tilde{\rho}^\dagger \tilde{\rho}}\right]/2. \tag{B.6}$$

The set of t_i needed to evaluate the function L a first time is obtained by the inversion of Eq. (B.3). Note that ρ_0 is already a physical matrix but is optimized in the Frobenius norm which has no direct physical meaning.

Bibliography

[Abdumalikov10] A. A. Abdumalikov, O. Astafiev, A. M. Zagoskin *et al.* 'Electromagnetically induced transparency on a single artificial atom'. *arXiv.org*, **1004.2306** (2010).

[Abragam61] A. Abragam. Principles of Nuclear Magnetism. Oxford University Press (1961).

[Aharonov88] Y. Aharonov, D. Z. Albert and L. Vaidman. 'How the result of a measurement of a component of the spin of a spin-1/2 particle can turn out to be 100'. *Phys. Rev. Lett.*, **60**(14):1351 (1988).

[Allen87] L. Allen and J. H. Eberly. Optical resonance and two-level atoms. Dover (1987).

[Ansmann09] M. Ansmann, H. Wang, R. C. Bialczak *et al.* 'Violation of Bell's inequality in Josephson phase qubits'. *Nature*, **461**(7263):504 (2009).

[Ansoft-Corp.05] Ansoft-Corp. 'Maxwell-10'. www.ansoft.com (2005).

[Armstrong78] G. Armstrong, A. S. Greenberg and J. R. Sites. 'Very low temperature thermal conductivity and optical properties of Stycast 1266 epoxy'. *Rev. Sci. Instrum.*, **49**(3):345 (1978).

[Aspect82] A. Aspect, J. Dalibard and G. Roger. 'Experimental Test of Bell's Inequalities Using Time- Varying Analyzers'. *Phys. Rev. Lett.*, **49**(25):1804 (1982).

[Astafiev04] O. Astafiev, Y. A. Pashkin, T. Yamamoto *et al.* 'Single-shot measurement of the Josephson charge qubit'. *Phys. Rev. B*, **69**(18):180507 (2004).

[Astafiev07] O. Astafiev, K. Inomata, A. O. Niskanen *et al.* 'Single artificial-atom lasing'. *Nature*, **449**(7162):588 (2007).

[AWR-Corp.06] AWR-Corp. 'Microwave Office 2006'. www.awrcorp.com (2006).

BIBLIOGRAPHY

[Barends08] R. Barends, H. L. Hortensius, T. Zijlstra et al. 'Contribution of dielectrics to frequency and noise of NbTiN superconducting resonators'. *Appl. Phys. Lett.*, **92**(22):223502 (2008).

[Barends10] R. Barends, N. Vercruyssen, A. Endo et al. 'Minimal resonator loss for circuit quantum electrodynamics'. *arXiv.org*, **1005.0408** (2010).

[Baur09] M. Baur, S. Filipp, R. Bianchetti et al. 'Measurement of Autler-Townes and Mollow Transitions in a Strongly Driven Superconducting Qubit'. *Phys. Rev. Lett.*, **102**(24):243602 (2009).

[Bell64] J. S. Bell. 'On the Einstein Podolsky Rosen Paradox'. *Physics*, **1**:195 (1964).

[Bennett82] C. H. Bennett. 'The thermodynamics of computation — a review'. *Internat. J. Theoret. Phys.*, **21**(12):905 (1982).

[Bennett84] C. H. Bennett and G. Brassard. 'Quantum cryptography: public key distribution and coin tossing'. *International Conference on Computers, Systems and Signal Processing*, (Bangalore, India):175 (1984).

[Bennett93] C. H. Bennett, G. Brassard, C. Crépeau et al. 'Teleporting an unknown quantum state via dual classical and Einstein-Podolsky-Rosen channels'. *Phys. Rev. Lett.*, **70**(13):1895 (1993).

[Berezovsky06] J. Berezovsky, M. H. Mikkelsen, O. Gywat et al. 'Nondestructive Optical Measurements of a Single Electron Spin in a Quantum Dot'. *Science*, **314**(5807):1916 (2006).

[Bergeal10] N. Bergeal, F. Schackert, M. Metcalfe et al. 'Phase-preserving amplification near the quantum limit with a Josephson ring modulator'. *Nature*, **465**(7294):64– (2010).

[Bhaumik75] K. Bhaumik and B. Dutta-Roy. 'The classical nonlinear oscillator and the coherent state'. *J. Math. Phys.*, **16**(5):1131 (1975).

[Bialczak10] R. C. Bialczak, M. Ansmann, M. Hofheinz et al. 'Quantum process tomography of a universal entangling gate implemented with Josephson phase qubits'. *Nat. Phys.* (2010).

[Bianchetti09] R. Bianchetti, S. Filipp, M. Baur et al. 'Dynamics of dispersive single-qubit readout in circuit quantum electrodynamics'. *Phys. Rev. A*, **80**(4):043840 (2009).

[Bianchetti10] R. Bianchetti, S. Filipp, M. Baur et al. 'Control and Tomography of a Three Level Superconducting Artificial Atom'. *arXiv.org*, **1004.5504** (2010).

BIBLIOGRAPHY

[Birnbaum05] K. M. Birnbaum, A. Boca, R. Miller et al. 'Photon blockade in an optical cavity with one trapped atom'. *Nature*, **436**(7047):87 (2005).

[Bishop09a] L. S. Bishop, J. M. Chow, J. Koch et al. 'Nonlinear response of the vacuum Rabi resonance'. *Nat. Phys.*, **5**(2):105 (2009).

[Bishop09b] L. S. Bishop, L. Tornberg, D. Price et al. 'Proposal for generating and detecting multi-qubit GHZ states in circuit QED'. *New J. Phys.*, **11**(7):073040 (2009).

[Bishop10] L. S. Bishop, E. Ginossar and S. M. Girvin. 'Response of the Strongly-Driven Jaynes-Cummings Oscillator'. *arXiv.org*, **1005.0377** (2010).

[Blais04] A. Blais, R.-S. Huang, A. Wallraff et al. 'Cavity quantum electrodynamics for superconducting electrical circuits: An architecture for quantum computation'. *Phys. Rev. A*, **69**(6):062320 (2004).

[Bloch08] I. Bloch, J. Dalibard and W. Zwerger. 'Many-body physics with ultracold gases'. *Rev. Mod. Phys.*, **80**(3):885 (2008).

[Boissonneault07] M. Boissonneault. 'Effets non-linéaires et qualité de la mesure en électrodynamique quantique en circuit' (2007). Master Thesis.

[Boissonneault08] M. Boissonneault, J. M. Gambetta and A. Blais. 'Nonlinear dispersive regime of cavity QED: The dressed dephasing model'. *Phys. Rev. A*, **77**(6):305 (2008).

[Boissonneault09] M. Boissonneault, J. M. Gambetta and A. Blais. 'Dispersive regime of circuit QED: Photon-dependent qubit dephasing and relaxation rates'. *Phys. Rev. A*, **79**(1):013819 (2009).

[Boissonneault10] M. Boissonneault, J. M. Gambetta and A. Blais. 'Improved Superconducting Qubit Readout by Qubit-Induced Nonlinearities'. *arXiv.org*, **1005.0004** (2010).

[Bose87] S. K. Bose and U. B. Dubey. 'Application of Coherent States to Anharmonic, Time-Dependent and Damped Oscillator Systems'. *Fortschr. Phys.*, **35**(10):675 (1987).

[Bozyigit10] D. Bozyigit, C. Lang, L. Steffen et al. 'Measurements of the Correlation Function of a Microwave Frequency Single Photon Source'. *arXiv.org*, **1002.3738** (2010).

[Brune96] M. Brune, F. Schmidt-Kaler, A. Maali et al. 'Quantum Rabi Oscillation: A Direct Test of Field Quantization in a Cavity'. *Phys. Rev. Lett.*, **76**(11):1800 (1996).

BIBLIOGRAPHY

[Buluta09] I. Buluta and F. Nori. 'Quantum Simulators'. *Science*, **326**(5949):108 (2009).

[Castellanos08] M. A. Castellanos, K. D. Irwin, G. C. Hilton *et al*. 'Amplification and squeezing of quantum noise with a tunable Josephson metamaterial'. *Nat. Phys.*, **4**(12):929 (2008).

[Caves82] C. M. Caves. 'Quantum limits on noise in linear amplifiers'. *Phys. Rev. D*, **26**(8):1817 (1982).

[Cerf02] N. J. Cerf, M. Bourennane, A. Karlsson *et al*. 'Security of Quantum Key Distribution Using d-Level Systems'. *Phys. Rev. Lett.*, **88**(12):127902 (2002).

[Cerletti05] V. Cerletti, W. A. Coish, O. Gywat *et al*. 'Recipes for spin-based quantum computing'. *Nanotechnology*, **16**(4):R27 (2005).

[Chen08] W. Chen, D. A. Bennett, V. Patel *et al*. 'Substrate and process dependent losses in superconducting thin film resonators'. *Supercond. Sci. Technol.*, **21**(7):075013 (2008).

[Childs10] A. M. Childs and W. van Dam. 'Quantum algorithms for algebraic problems'. *Rev. Mod. Phys.*, **82**(1):1 (2010).

[Chiorescu03] I. Chiorescu, Y. Nakamura, C. J. P. M. Harmans *et al*. 'Coherent quantum dynamics of a superconducting flux qubit'. *Science*, **299**(5614):1869 (2003).

[Chiorescu04] I. Chiorescu, P. Bertet, K. Semba *et al*. 'Coherent dynamics of a flux qubit coupled to a harmonic oscillator'. *Nature*, **431**(7005):159 (2004).

[Chow09a] J. M. Chow, L. DiCarlo, J. M. Gambetta *et al*. 'Entanglement Metrology Using a Joint Readout of Superconducting Qubits'. *arXiv.org*, **0908.1955** (2009).

[Chow09b] J. M. Chow, J. M. Gambetta, L. Tornberg *et al*. 'Randomized Benchmarking and Process Tomography for Gate Errors in a Solid-State Qubit'. *Phys. Rev. Lett.*, **102**(9):090502 (2009).

[Clarke08] J. Clarke and F. K. Wilhelm. 'Superconducting quantum bits'. *Nature*, **453**(7198):1031 (2008).

[Claudon04] J. Claudon, F. Balestro, F. W. J. Hekking *et al*. 'Coherent Oscillations in a Superconducting Multilevel Quantum System'. *Phys. Rev. Lett.*, **93**(18):187003 (2004).

[Clerk10] A. A. Clerk, M. H. Devoret, S. M. Girvin *et al*. 'Introduction to quantum noise, measurement, and amplification'. *Rev. Mod. Phys.*, **82**(2):1155 (2010).

BIBLIOGRAPHY

[Constantin09] M. Constantin, C. C. Yu and J. M. Martinis. 'Saturation of two-level systems and charge noise in Josephson junction qubits'. *Phys. Rev. B*, **79**(9):094520 (2009).

[Cooper04] K. B. Cooper, M. Steffen, R. McDermott *et al.* 'Observation of Quantum Oscillations between a Josephson Phase Qubit and a Microscopic Resonator Using Fast Readout'. *Phys. Rev. Lett.*, **93**(18):180401 (2004).

[Dell'Anno06] F. Dell'Anno, S. De Siena and F. Illuminati. 'Multiphoton quantum optics and quantum state engineering'. *Phys. Rep.*, **428**(2-3):53 (2006).

[Deutsch85] D. Deutsch. 'Quantum Theory, the Church-Turing Principle and the Universal Quantum Computer'. *P. Roy. Soc. Lond. A Mat*, **400**(1818):97 (1985).

[Devoret00] M. H. Devoret and R. J. Schoelkopf. 'Amplifying quantum signals with the single-electron transistor'. *Nature*, **406**(6799):1039 (2000).

[DiCarlo09] L. DiCarlo, J. M. Chow, J. M. Gambetta *et al.* 'Demonstration of two-qubit algorithms with a superconducting quantum processor'. *Nature*, **460**(7252):240 (2009).

[DiCarlo10] L. DiCarlo, M. D. Reed, L. Sun *et al.* 'Preparation and Measurement of Three-Qubit Entanglement in a Superconducting Circuit'. *arXiv.org*, **1004.4324** (2010).

[DiVincenzo97] D. P. DiVincenzo. 'Topics in Quantum Computers'. *Mesoscopic Electron Transport (NATO Advanced Study Institute, Series E: Applied Sciences)*, **345** (1997).

[DiVincenzo00] D. P. DiVincenzo. 'The Physical Implementation of Quantum Computation'. *Fortschr. Phys.*, **48**(9-11):771 (2000).

[Duan10] L. M. Duan and C. Monroe. 'Colloquium: Quantum networks with trapped ions'. *Rev. Mod. Phys.*, **82**(2):1209 (2010).

[Duffing18] G. Duffing. 'Erzwungene Schwingungen bei veränderlicher Eigenfrequenz'. *F. Vieweg*, **41/42**:134 (1918).

[Durt04] T. Durt, D. Kaszlikowski, J. L. Chen *et al.* 'Security of quantum key distributions with entangled qudits'. *Phys. Rev. A*, **69**(3):032313 (2004).

[Dutt07] M. V. G. Dutt, L. Childress, L. Jiang *et al.* 'Quantum Register Based on Individual Electronic and Nuclear Spin Qubits in Diamond'. *Science*, **316**(5829):1312 (2007).

BIBLIOGRAPHY

[Dutta08] S. K. Dutta, F. W. Strauch, R. M. Lewis et al. 'Multilevel effects in the Rabi oscillations of a Josephson phase qubit'. *Phys. Rev. B*, **78**(10):104510 (2008).

[Duty04] T. Duty, D. Gunnarsson, K. Bladh et al. 'Coherent dynamics of a Josephson charge qubit'. *Phys. Rev. B*, **69**(14):140503 (2004).

[Dykman05] M. I. Dykman and M. V. Fistul. 'Multiphoton antiresonance'. *Phys. Rev. B*, **71**(14):140508 (2005).

[Dykman07] M. I. Dykman. 'Critical exponents in metastable decay via quantum activation'. *Phys. Rev. E*, **75**(1):011101 (2007).

[Einstein35] A. Einstein, B. Podolsky and N. Rosen. 'Can Quantum-Mechanical Description of Physical Reality Be Considered Complete?' *Physical Review*, **47**(10):777 (1935).

[Ferrón10] A. Ferrón and D. Domínguez. 'Intrinsic leakage of the Josephson flux qubit and breakdown of the two-level approximation for strong driving'. *Phys. Rev. B*, **81**(10):104505 (2010).

[Feynman82] R. P. Feynman. 'Simulating physics with computers'. *Int. J. Theor. Phys.*, **21**(6):467 (1982).

[Filipp09] S. Filipp, P. Maurer, P. J. Leek et al. 'Two-Qubit State Tomography Using a Joint Dispersive Readout'. *Phys. Rev. Lett.*, **102**(20):200402 (2009).

[Fink08] J. M. Fink, M. Göppl, M. Baur et al. 'Climbing the Jaynes-Cummings ladder and observing its nonlinearity in a cavity QED system'. *Nature*, **454**(7202):315 (2008).

[Fink09] J. M. Fink, R. Bianchetti, M. Baur et al. 'Dressed Collective Qubit States and the Tavis-Cummings Model in Circuit QED'. *Phys. Rev. Lett.*, **103**(8):083601 (2009).

[Fink10] J. M. Fink, L. Steffen, P. Studer et al. 'Quantum-to-Classical Transition in Cavity Quantum Electrodynamics (QED)'. *arXiv.org*, **1003.1161** (2010).

[Fragner08] A. Fragner, M. Göppl, J. M. Fink et al. 'Resolving Vacuum Fluctuations in an Electrical Circuit by Measuring the Lamb Shift'. *Science*, **322**(5906):1357 (2008).

[Fukushima97] A. Fukushima, A. Sato, A. Iwasa et al. 'Attenuation of microwave filters for single-electron tunneling experiments'. *Instrumentation and Measurement, IEEE Transactions on*, **46**(2):289 (1997).

BIBLIOGRAPHY

[Gaebel06]   T. Gaebel, M. Domhan, I. Popa et al. 'Room-temperature coherent coupling of single spins in diamond'. *Nat. Phys.*, **2**(6):408 (2006).

[Gambetta06]   J. Gambetta, A. Blais, D. I. Schuster et al. 'Qubit-photon interactions in a cavity: Measurement-induced dephasing and number splitting'. *Phys. Rev. A*, **74**(4):042318 (2006).

[Gambetta07]   J. Gambetta, W. A. Braff, A. Wallraff et al. 'Protocols for optimal readout of qubits using a continuous quantum nondemolition measurement'. *Phys. Rev. A*, **76**(1):012325 (2007).

[Gambetta08]   J. Gambetta, A. Blais, M. Boissonneault et al. 'Quantum trajectory approach to circuit QED: Quantum jumps and the Zeno effect'. *Phys. Rev. A*, **77**(1):012112 (2008).

[Gao06]   J. Gao, B. A. Mazin, M. Daal et al. 'Power dependence of phase noise in microwave kinetic inductance detectors'. *Millimeter and Submillimeter Detectors and Instrumentation for Astronomy III*, **6275**(1):627509 (2006).

[Gao07]   J. Gao, J. Zmuidzinas, B. A. Mazin et al. 'Noise properties of superconducting coplanar waveguide microwave resonators'. *Appl. Phys. Lett.*, **90**(10):102507 (2007).

[Gao08a]   J. Gao, M. Daal, J. M. Martinis et al. 'A semiempirical model for two-level system noise in superconducting microresonators'. *Appl. Phys. Lett.*, **92**(21):212504 (2008).

[Gao08b]   J. Gao, M. Daal, A. Vayonakis et al. 'Experimental evidence for a surface distribution of two-level systems in superconducting lithographed microwave resonators'. *Appl. Phys. Lett.*, **92**(15):152505 (2008).

[Gardiner85]   C. W. Gardiner and M. J. Collett. 'Input and output in damped quantum systems: Quantum stochastic differential equations and the master equation'. *Phys. Rev. A*, **31**(6):3761 (1985).

[Giovannetti06]   V. Giovannetti, S. Lloyd and L. Maccone. 'Quantum Metrology'. *Phys. Rev. Lett.*, **96**(1):010401 (2006).

[Girvin09]   S. M. Girvin, M. H. Devoret and R. J. Schoelkopf. 'Circuit QED and engineering charge-based superconducting qubits'. *Phys. Scr.*, **2009**(T137):014012 (2009).

[Gisin02]   N. Gisin, G. Ribordy, W. Tittel et al. 'Quantum cryptography'. *Rev. Mod. Phys.*, **74**(1):145 (2002).

BIBLIOGRAPHY

[Göppl08] M. Göppl, A. Fragner, M. Baur et al. 'Coplanar Waveguide Resonators for Circuit Quantum Electrodynamics'. *J. Appl. Phys.*, **104**(11):113904 (2008).

[Göppl09] M. Göppl. Engineering Quantum Electronic Chips - Realization and Characterization of Circuit Quantum Electrodynamics Systems. Ph.D. thesis, ETH Zurich (2009).

[Goy83] P. Goy, J. M. Raimond, M. Gross et al. 'Observation of Cavity-Enhanced Single-Atom Spontaneous Emission'. *Phys. Rev. Lett.*, **50**(24):1903 (1983).

[Grajcar04] M. Grajcar, A. Izmalkov, E. Il'ichev et al. 'Low-frequency measurement of the tunneling amplitude in a flux qubit'. *Phys. Rev. B*, **69**(6):060501 (2004).

[Häffner08] H. Häffner, C. F. Roos and R. Blatt. 'Quantum computing with trapped ions'. *Physics Reports*, **469**(4):155 (2008).

[Hammer07] G. Hammer, S. Wuensch, M. Roesch et al. 'Superconducting coplanar waveguide resonators for detector applications'. *Supercond. Sci. Technol.*, **20**(11):S408 (2007).

[Hammerer10] K. Hammerer, A. S. Sørensen and E. S. Polzik. 'Quantum interface between light and atomic ensembles'. *Rev. Mod. Phys.*, **82**(2):1041 (2010).

[Hanson07] R. Hanson, L. P. Kouwenhoven, J. R. Petta et al. 'Spins in few-electron quantum dots'. *Rev. Mod. Phys.*, **79**(4):1217 (2007).

[Helmer09a] F. Helmer, M. Mariantoni, A. G. Fowler et al. 'Cavity grid for scalable quantum computation with superconducting circuits'. *Europhys. Lett.*, **85**(5):50007 (2009).

[Helmer09b] F. Helmer and F. Marquardt. 'Measurement-based synthesis of multiqubit entangled states in superconducting cavity QED'. *Phys. Rev. A*, **79**(5):052328 (2009).

[Higham86] N. J. Higham. 'Computing the Polar Decomposition-with Applications'. *SIAM Journal on Scientific and Statistical Computing*, **7**(4):1160 (1986).

[Higham88] N. J. Higham. 'Computing a nearest symmetric positive semidefinite matrix'. *Linear Algebra Appl.*, **103**:103 (1988).

[Hofheinz08] M. Hofheinz, E. M. Weig, M. Ansmann et al. 'Generation of Fock states in a superconducting quantum circuit'. *Nature*, **454**(7202):310 (2008).

[Hofheinz09] M. Hofheinz, H. Wang, M. Ansmann et al. 'Synthesizing arbitrary quantum states in a superconducting resonator'. *Nature*, **459**(7246):546 (2009).

BIBLIOGRAPHY

[Holmes76] P. J. Holmes and D. A. Rand. 'The bifurcations of duffing's equation: An application of catastrophe theory'. *J. Sound Vibrat.*, **44**(2):237 (1976).

[Horodecki09] R. Horodecki, P. Horodecki, M. Horodecki *et al.* 'Quantum entanglement'. *Rev. Mod. Phys.*, **81**(2):865 (2009).

[Houck07] A. A. Houck, D. I. Schuster, J. M. Gambetta *et al.* 'Generating single microwave photons in a circuit'. *Nature*, **449**(7160):328 (2007).

[Houck08] A. A. Houck, J. A. Schreier, B. R. Johnson *et al.* 'Controlling the Spontaneous Emission of a Superconducting Transmon Qubit'. *Phys. Rev. Lett.*, **101**(8):080502 (2008).

[IEE00] IEEE 100 The Authoritative Dictionary of IEEE Standards Terms Seventh Edition. IEEE (2000).

[Inoue09] R. Inoue, T. Yonehara, Y. Miyamoto *et al.* 'Measuring Qutrit-Qutrit Entanglement of Orbital Angular Momentum States of an Atomic Ensemble and a Photon'. *Phys. Rev. Lett.*, **103**(11):110503 (2009).

[Ithier05] G. Ithier, E. Collin, P. Joyez *et al.* 'Decoherence in a superconducting quantum bit circuit'. *Phys. Rev. B*, **72**:134519 (2005).

[James01] D. F. V. James, P. G. Kwiat, W. J. Munro *et al.* 'Measurement of qubits'. *Phys. Rev. A*, **64**(5):052312 (2001).

[Jaynes63] E. T. Jaynes and F. W. Cummings. 'Comparison of quantum and semiclassical radiation theories with application to the beam maser'. **51**(1):89 (1963).

[Jirari05] H. Jirari and W. Pötz. 'Optimal coherent control of dissipative N-level systems'. *Phys. Rev. A*, **72**(1):013409 (2005).

[Jordan10] A. N. Jordan and A. N. Korotkov. 'Uncollapsing the wavefunction by undoing quantum measurements'. *Contemporary Physics*, **51**(2):125 (2010).

[Kaszlikowski00] D. Kaszlikowski, P. Gnaciński, M. Żukowski *et al.* 'Violations of Local Realism by Two Entangled N-Dimensional Systems Are Stronger than for Two Qubits'. *Phys. Rev. Lett.*, **85**(21):4418 (2000).

[Katz06] N. Katz, M. Ansmann, R. C. Bialczak *et al.* 'Coherent state evolution in a superconducting qubit from partial-collapse measurement'. *Science*, **312**(5779):1498 (2006).

[Katz08] N. Katz, M. Neeley, M. Ansmann *et al.* 'Reversal of the Weak Measurement of a Quantum State in a Superconducting Phase Qubit'. *Phys. Rev. Lett.*, **101**(20):200401 (2008).

BIBLIOGRAPHY

[Kerr99] A. R. Kerr. 'Suggestions for revised definitions of noise quantities, including quantum effects'. *Microwave Theory and Techniques, IEEE Transactions on*, **47**(3):325 (1999).

[Kitazawa86] T. Kitazawa and Y. Hayashi. 'Quasistatic characteristics of a coplanar waveguide with thick metal coating'. *Microwaves, Antennas and Propagation, IEE Proceedings H*, **133**(1):18 (1986).

[Knill00] E. Knill, R. Laflamme and L. Viola. 'Theory of Quantum Error Correction for General Noise'. *Phys. Rev. Lett.*, **84**(11):2525 (2000).

[Koch07] J. Koch, T. M. Yu, J. Gambetta *et al.* 'Charge-insensitive qubit design derived from the Cooper pair box'. *Phys. Rev. A*, **76**(4):042319 (2007).

[Kok07] P. Kok, W. J. Munro, K. Nemoto *et al.* 'Linear optical quantum computing with photonic qubits'. *Rev. Mod. Phys.*, **79**(1):135 (2007).

[Krivoshlykov82] S. G. Krivoshlykov, V. I. Man'ko and I. N. Sissakian. 'Coherent state evolution for the quantum anharmonic oscillator'. *Phys. Lett. A*, **90**(4):65 (1982).

[Kumar08] S. Kumar, J. Gao, J. Zmuidzinas *et al.* 'Temperature dependence of the frequency and noise of superconducting coplanar waveguide resonators'. *Appl. Phys. Lett.*, **92**(12):123503 (2008).

[Lake-Shore04] Lake-Shore. 'Temperature and measurement catalog'. www.lakeshore.com (2004).

[Landau76] L. D. Landau and E. M. Lifshitz. Mechanics, volume 1. Elsevier, 3 edition (1976).

[Lang09] C. Lang. 'Read-Out Strategies for Multi-Qubit States in Circuit Quantum Electrodynamics' (2009). Diploma Thesis.

[Lanyon09] B. P. Lanyon, M. Barbieri, M. P. Almeida *et al.* 'Simplifying quantum logic using higher-dimensional Hilbert spaces'. *Nat. Phys.*, **5**(2):134 (2009).

[Leek07] P. J. Leek, J. M. Fink, A. Blais *et al.* 'Observation of Berry's Phase in a Solid-State Qubit'. *Science*, **318**(5858):1889 (2007).

[Leek09] P. J. Leek, S. Filipp, P. Maurer *et al.* 'Using sideband transitions for two-qubit operations in superconducting circuits'. *Phys. Rev. B*, **79**(18):180511 (2009).

[Leek10] P. J. Leek, M. Baur, J. M. Fink *et al.* 'Cavity Quantum Electrodynamics with Separate Photon Storage and Qubit Readout Modes'. *Phys. Rev. Lett.*, **104**(10):100504 (2010).

BIBLIOGRAPHY

[Leibfried03]   D. Leibfried, R. Blatt, C. Monroe et al. 'Quantum dynamics of single trapped ions'. *Rev. Mod. Phys.*, **75**(1):281 (2003).

[Leong02]   K. Leong and J. Mazierska. 'Precise measurements of the Q factor of dielectric resonators in the transmission mode-accounting for noise, crosstalk, delay of uncalibrated lines, coupling loss, and coupling reactance'. *Microwave Theory and Techniques, IEEE Transactions on*, **50**(9):2115 (2002).

[Lewenstein07]   M. Lewenstein, A. Sanpera, V. Ahufinger et al. 'Ultracold atomic gases in optical lattices: mimicking condensed matter physics and beyond'. *Advances in Physics*, **56**(2):243 (2007).

[Lindblad76]   G. Lindblad. 'On the Generators of Quantum Dynamical Semigroups'. *Commun. Math. Phys.*, **48**(2):119 (1976).

[Lloyd96]   S. Lloyd. 'Universal Quantum Simulators'. *Science*, **273**(5278):1073 (1996).

[Lucero08]   E. Lucero, M. Hofheinz, M. Ansmann et al. 'High-Fidelity Gates in a Single Josephson Qubit'. *Phys. Rev. Lett.*, **100**(24):247001 (2008).

[Lukashenko08]   A. Lukashenko and A. V. Ustinov. 'Improved powder filters for qubit measurements'. *Rev. Sci. Instrum.*, **79**(1):014701 (2008).

[Lupascu04]   A. Lupascu, C. J. M. Verwijs, R. N. Schouten et al. 'Nondestructive Readout for a Superconducting Flux Qubit'. *Phys. Rev. Lett.*, **93**(17):177006 (2004).

[Lupascu05]   A. Lupascu, C. J. P. M. Harmans and J. E. Mooij. 'Quantum state detection of a superconducting flux qubit using a dc-SQUID in the inductive mode'. *Phys. Rev. B*, **71**(18):184506 (2005).

[Lupascu06]   A. Lupascu, E. F. C. Driessen, L. Roschier et al. 'High-Contrast Dispersive Readout of a Superconducting Flux Qubit Using a Nonlinear Resonator'. *Phys. Rev. Lett.*, **96**(12):127003 (2006).

[Lupascu07]   A. Lupascu, S. Saito, T. Picot et al. 'Quantum non-demolition measurement of a superconducting two-level system'. *Nat. Phys.*, **3**(2):119 (2007).

[Mair01]   A. Mair, A. Vaziri, G. Weihs et al. 'Entanglement of the orbital angular momentum states of photons'. *Nature*, **412**(6844):313 (2001).

[Majer07]   J. Majer, J. M. Chow, J. M. Gambetta et al. 'Coupling superconducting qubits via a cavity bus'. *Nature*, **449**(7161):443 (2007).

[Makhlin01]   Y. Makhlin, G. Schön and A. Shnirman. 'Quantum-state engineering with Josephson-junction devices'. *Rev. Mod. Phys.*, **73**(2):357 (2001).

BIBLIOGRAPHY

[Mallet09] F. Mallet, F. R. Ong, A. Palacios-Laloy et al. 'Single-shot qubit readout in circuit quantum electrodynamics'. Nat. Phys., **5**(11):791 (2009).

[Marthaler06] M. Marthaler and M. I. Dykman. 'Switching via quantum activation: A parametrically modulated oscillator'. Phys. Rev. A, **73**(4):042108 (2006).

[Martinis87] J. M. Martinis, M. H. Devoret and J. Clarke. 'Experimental tests for the quantum behavior of a macroscopic degree of freedom: The phase difference across a Josephson junction'. Phys. Rev. B, **35**(10):4682 (1987).

[Martinis02] J. M. Martinis, S. Nam, J. Aumentado et al. 'Rabi oscillations in a large Josephson-junction qubit'. Phys. Rev. Lett., **89**(11):117901 (2002).

[Martinis05] J. M. Martinis, K. B. Cooper, R. McDermott et al. 'Decoherence in Josephson Qubits from Dielectric Loss'. Phys. Rev. Lett., **95**(21):210503 (2005).

[Martinis09] J. M. Martinis, M. Ansmann and J. Aumentado. 'Energy Decay in Superconducting Josephson-Junction Qubits from Nonequilibrium Quasiparticle Excitations'. Phys. Rev. Lett., **103**(9):097002 (2009).

[Mazin02] B. A. Mazin, P. K. Day, H. G. LeDuc et al. 'Superconducting Kinetic Inductance Photon Detectors'. 2002 Proc. SPIE, **4849**:283 (2002).

[McDermott05] R. McDermott, R. W. Simmonds, M. Steffen et al. 'Simultaneous state measurement of coupled Josephson phase qubits'. Science, **307**(5713):1299 (2005).

[McKinstry89] K. D. McKinstry and C. E. Patton. 'Methods for determination of microwave cavity quality factors from equivalent electronic circuit models'. Rev. Sci. Instrum., **60**(3):439 (1989).

[Molina-Terriza04] G. Molina-Terriza, A. Vaziri, J. Řeháček et al. 'Triggered Qutrits for Quantum Communication Protocols'. Phys. Rev. Lett., **92**(16):167903 (2004).

[Motzoi] F. Motzoi, J. Gambetta and F. Wilhelm. In preparation.

[Motzoi09] F. Motzoi, J. M. Gambetta, P. Rebentrost et al. 'Simple Pulses for Elimination of Leakage in Weakly Nonlinear Qubits'. Phys. Rev. Lett., **103**(11):110501 (2009).

[Murali04] K. V. R. M. Murali, Z. Dutton, W. D. Oliver et al. 'Probing Decoherence with Electromagnetically Induced Transparency in Superconductive Quantum Circuits'. Phys. Rev. Lett., **93**(8):087003 (2004).

[Nakamura99] Y. Nakamura, Y. A. Pashkin and J. S. Tsai. 'Coherent control of macroscopic quantum states in a single-Cooper-pair box'. Nature, **398**(6730):786 (1999).

BIBLIOGRAPHY

[Nakamura02] Y. Nakamura, Y. A. Pashkin, T. Yamamoto et al. 'Charge Echo in a Cooper-Pair Box'. *Phys. Rev. Lett.*, **88**(4):047901 (2002).

[Neeley09] M. Neeley, M. Ansmann, R. C. Bialczak et al. 'Emulation of a Quantum Spin with a Superconducting Phase Qudit'. *Science*, **325**(5941):722 (2009).

[Nielsen00] M. A. Nielsen and I. L. Chuang. Quantum Computation and Quantum Information. Cambridge Univertity Press (2000).

[O'Connell08] A. D. O'Connell, M. Ansmann, R. C. Bialczak et al. 'Microwave dielectric loss at single photon energies and millikelvin temperatures'. *Appl. Phys. Lett.*, **92**(11):112903 (2008).

[Pashkin09] Y. A. Pashkin, O. Astafiev, T. Yamamoto et al. 'Josephson charge qubits: a brief review'. *Quantum Inf. Process.*, **8**:55 (2009).

[Plantenberg07] J. H. Plantenberg, P. C. de Groot, C. J. P. M. Harmans et al. 'Demonstration of controlled-NOT quantum gates on a pair of superconducting quantum bits'. *Nature*, **447**(7146):836 (2007).

[Pobell06] F. Pobell. Matter and Methods at Low Temperatures. Springer (2006).

[Porch95] A. Porch, M. J. Lancaster and R. G. Humphreys. 'The Coplanar Resonator Technique for Determining the Surface Impedance of YBA2CU3O7-Delta Thin-Films'. *IEEE T. Microw. Theory.*, **43**(2):306 (1995).

[Pozar90] D. N. Pozar. Microwave Engineering. Wiley Inter-Science (1990).

[Raimond01] J. M. Raimond and S. Brune, M.and Haroche. 'Colloquium: Manipulating quantum entanglement with atoms and photons in a cavity'. *Rev. Mod. Phys.*, **73**(3):565 (2001).

[Rebentrost09] P. Rebentrost and F. K. Wilhelm. 'Optimal control of a leaking qubit'. *Phys. Rev. B*, **79**(6):060507 (2009).

[Reed10a] M. D. Reed, L. Di Carlo, B. R. Johnson et al. 'High Fidelity Readout in Circuit Quantum Electrodynamics Using the Jaynes-Cummings Nonlinearity'. *arXiv.org*, **1004.4323** (2010).

[Reed10b] M. D. Reed, B. R. Johnson, A. A. Houck et al. 'Fast Reset and Suppressing Spontaneous Emission of a Superconducting Qubit'. *arXiv.org*, **1003.0142** (2010).

[Reithmaier04] J. P. Reithmaier, G. Sek, A. Loffler et al. 'Strong coupling in a single quantum dot-semiconductor microcavity system'. *Nature*, **432**(7014):197 (2004).

BIBLIOGRAPHY

[Rempe87] G. Rempe, H. Walther and N. Klein. 'Observation of quantum collapse and revival in a one-atom maser'. *Phys. Rev. Lett.*, **58**(4):353 (1987).

[Rozanov81] N. N. Rozanov and V. A. Smirnov. 'Resonance excitation and hysteresis in a quantum anharmonic oscillator'. *JETP Lett.*, **33**:488 (1981).

[Safaei09] S. Safaei, S. Montangero, F. Taddei *et al.* 'Optimized single-qubit gates for Josephson phase qubits'. *Phys. Rev. B*, **79**(6):064524 (2009).

[Scarani09] V. Scarani, H. Bechmann-Pasquinucci, N. J. Cerf *et al.* 'The security of practical quantum key distribution'. *Rev. Mod. Phys.*, **81**(3):1301 (2009).

[Schreier08] J. A. Schreier, A. A. Houck, J. Koch *et al.* 'Suppressing charge noise decoherence in superconducting charge qubits'. *Phys. Rev. B*, **77**(18):180502(R) (2008).

[Schrödinger35] E. Schrödinger. 'Die gegenwärtige Situation in der Quantenmechanik'. *Naturwissenschaften*, **23**(48):807 (1935).

[Schuster05] D. I. Schuster, A. Wallraff, A. Blais *et al.* 'ac Stark shift and dephasing of a superconducting qubit strongly coupled to a cavity field'. *Phys. Rev. Lett.*, **94**(12):123602 (2005).

[Schuster07a] D. I. Schuster. Circuit Quantum Electrodynamics. Ph.D. thesis, Yale University (2007).

[Schuster07b] D. I. Schuster, A. A. Houck, J. A. Schreier *et al.* 'Resolving photon number states in a superconducting circuit'. *Nature*, **445**(7127):515 (2007).

[Serban07] I. Serban and F. K. Wilhelm. 'Dynamical Tunneling in Macroscopic Systems'. *Phys. Rev. Lett.*, **99**(13):137001 (2007).

[Serban10] I. Serban, M. I. Dykman and F. K. Wilhelm. 'Relaxation of a qubit measured by a driven Duffing oscillator'. *Phys. Rev. A*, **81**(2):022305 (2010).

[Shnirman05] A. Shnirman, G. Schön, I. Martin *et al.* 'Low- and High-Frequency Noise from Coherent Two-Level Systems'. *Phys. Rev. Lett.*, **94**(12):127002 (2005).

[Siddiqi04] I. Siddiqi, R. Vijay, F. Pierre *et al.* 'RF-Driven Josephson Bifurcation Amplifier for Quantum Measurement'. *Phys. Rev. Lett.*, **93**(20):207002 (2004).

[Siddiqi06] I. Siddiqi, R. Vijay, M. Metcalfe *et al.* 'Dispersive measurements of superconducting qubit coherence with a fast latching readout'. *Phys. Rev. B*, **73**(5):054510 (2006).

BIBLIOGRAPHY

[Sillanpää05]   M. A. Sillanpää, T. Lehtinen, A. Paila et al. 'Direct Observation of Josephson Capacitance'. *Phys. Rev. Lett.*, **95**(20):206806 (2005).

[Sillanpää09]   M. A. Sillanpää, J. Li, K. Cicak et al. 'Autler-Townes Effect in a Superconducting Three-Level System'. *Phys. Rev. Lett.*, **103**(19):193601 (2009).

[Simmonds04]   R. W. Simmonds, K. M. Lang, D. A. Hite et al. 'Decoherence in Josephson phase qubits from Junction resonators'. *Phys. Rev. Lett.*, **93**(7):077003 (2004).

[Simons01]   R. N. Simons. Coplanar waveguide circuits, components and systems. Wiley Inter-Science (2001).

[Skolnick04]   M. S. Skolnick and D. J. Mowbray. 'SELF-ASSEMBLED SEMICONDUCTOR QUANTUM DOTS: Fundamental Physics and Device Applications'. *Annu. Rev. Mater. Res.*, **34**(1):181 (2004).

[Spiller05]   T. P. Spiller, W. J. Munro, S. D. Barrett et al. 'An introduction to quantum information processing: applications and realizations'. *Contemporary Physics*, **46**(407):6 (2005).

[Steffen03]   M. Steffen, J. M. Martinis and I. L. Chuang. 'Accurate control of Josephson phase qubits'. *Phys. Rev. B*, **68**(22):224518 (2003).

[Thew02]   R. T. Thew, K. Nemoto, A. G. White et al. 'Qudit quantum-state tomography'. *Phys. Rev. A*, **66**(1):012303 (2002).

[Thew04]   R. T. Thew, A. Acín, H. Zbinden et al. 'Bell-Type Test of Energy-Time Entangled Qutrits'. *Phys. Rev. Lett.*, **93**(1):010503 (2004).

[Tinkham96]   M. Tinkham. Introduction to Superconductivity. McGraw-Hill International Editions (1996).

[Turing37]   A. M. Turing. 'On Computable Numbers, with an Application to the Entscheidungsproblem'. *Proc. London Math. Soc.*, **42**(1):230 (1937).

[Uzan03]   J. P. Uzan. 'The fundamental constants and their variation: observational and theoretical status'. *Rev. Mod. Phys.*, **75**(2):403 (2003).

[Vallone07]   G. Vallone, E. Pomarico, F. De Martini et al. 'Experimental realization of polarization qutrits from nonmaximally entangled states'. *Phys. Rev. A*, **76**(1):012319 (2007).

[vanderWal00]   C. H. van der Wal, A. C. J. ter Haar, F. K. Wilhelm et al. 'Quantum superposition of macroscopic persistent-current states'. *Science*, **290**(5492):773 (2000).

BIBLIOGRAPHY

[Vijay09] R. Vijay, M. H. Devoret and I. Siddiqi. 'Invited Review Article: The Josephson bifurcation amplifier'. *Rev. Sci. Instrum.*, **80**(11):111101 (2009).

[Vion02] D. Vion, A. Aassime, A. Cottet *et al.* 'Manipulating the quantum state of an electrical circuit'. *Science*, **296**(5569):886 (2002).

[Wallraff04] A. Wallraff, D. I. Schuster, A. Blais *et al.* 'Strong coupling of a single photon to a superconducting qubit using circuit quantum electrodynamics'. *Nature*, **431**(7005):162 (2004).

[Wallraff05] A. Wallraff, D. I. Schuster, A. Blais *et al.* 'Approaching Unit Visibility for Control of a Superconducting Qubit with Dispersive Readout'. *Phys. Rev. Lett.*, **95**(6):060501 (2005).

[Walls94] D. Walls and G. Milburn. Quantum optics. Spinger-Verlag, Berlin (1994).

[Walther06] H. Walther, B. T. H. Varcoe, B. G. Englert *et al.* 'Cavity quantum electrodynamics'. *Rep. Prog. Phys.*, **69**(5):1325 (2006).

[Wang08] H. Wang, M. Hofheinz, M. Ansmann *et al.* 'Measurement of the Decay of Fock States in a Superconducting Quantum Circuit'. *Phys. Rev. Lett.*, **101**(24):240401 (2008).

[Wang09a] H. Wang, M. Hofheinz, M. Ansmann *et al.* 'Decoherence Dynamics of Complex Photon States in a Superconducting Circuit'. *Phys. Rev. Lett.*, **103**(20):200404 (2009).

[Wang09b] H. Wang, M. Hofheinz, J. Wenner *et al.* 'Improving the coherence time of superconducting coplanar resonators'. *Appl. Phys. Lett.*, **95**(23):233508 (2009).

[Watanabe94] K. Watanabe, K. Yohida, T. Aoki *et al.* 'Kinetic Inductance of Superconducting Coplanar Wave-Guides'. *Jpn. J. Appl. Phys.*, **33**(10):5708 (1994).

[Wilson10] C. M. Wilson, G. Johansson, T. Duty *et al.* 'Dressed relaxation and dephasing in a strongly driven two-level system'. *Phys. Rev. B*, **81**(2):024520 (2010).

[Yamamoto99] Y. Yamamoto and A. Imamoglu. Mesoscopic Quantum Optics. Wiley (1999).

[Yamamoto03] T. Yamamoto, Y. A. Pashkin, O. Astafiev *et al.* 'Demonstration of conditional gate operation using superconducting charge qubits'. *Nature*, **425**(6961):941 (2003).

[Yoshida92] K. Yoshida, M. S. Hossain, T. Kisu *et al.* 'Modeling of Kinetic-Inductance Coplanar Stripline With Nbn Thin-Films'. *Jpn. J. Appl. Phys.*, **31**(12A):3844 (1992).

[Yoshie04]   T. Yoshie, A. Scherer, J. Hendrickson *et al.* 'Vacuum Rabi splitting with a single quantum dot in a photonic crystal nanocavity'. *Nature*, **432**:200 (2004).

[Zurek03]   W. H. Zurek. 'Decoherence, einselection, and the quantum origins of the classical'. *Rev. Mod. Phys.*, **75**(3):715 (2003).

Acknowledgements

This thesis could never ever have reached the present stage without the collaboration of the whole Quantum Device Lab team at ETH Zurich, which was of utmost importance in every single aspect of the work during the last four years. As group leader Andreas founded this effort, permitted me to join the team and escorted me during all this time, even below the surface of a dark lake in Lugano. I really want to thank him for all the thinks he allowed me to do, while constantly coaching and advising me. My gratitude goes to Prof. Dr. Alexey Ustinov from university of Karlsruhe for his co-supervision of my thesis, in the hope that some inspiration for his own related work arised from the reading of this thesis.

A big thank you goes to Hansruedi which has magic hands in fixing and manufacturing any kind of mechanical devices while at the same time giving irreplaceable advices over hiking destinations all around Switzerland. Gaby on her turn never managed to enter our offices without a big smile and always sorted out the messes from the "professional" shipping and delivery companies we deal with with a big laugh.

The constant presence of Pete in the case of any problem and the discussions on disparate thinks ranging from environmental issues to running advices passing over all possible details on circuit quantum electrodynamic experiments and theories merit my thankfulness. Without Stefan, three level state tomography in particular would still be a far goal and his very active and fruitful collaboration really justify my big gratefulness to him. He also rushed trough an early version of this thesis providing precious feedback and finding innumerable language mistakes.

In sparse order also many thanks to my early Phd. buddies Martin which grew the first samples at ETH and Josch with whom the first fridge was cabled and experiments performed.

Playing soccer with Johannes without loosing a leg and Diving disparate lakes with Martin and Matthias was great fun!

Sincere thanks go to Christian who, together with Deniz implemented our FPGA computer card used under others to perform single shots measurements. We had a lot of fruitful discussions over many possible measurement procedures and to be finally beaten by the Yale group in the development of a single shot readout was not too bad if one considers the good time we had and the amount of stuff both of us learned. I did not forget Matthias which is not only divemaster but also master of our Cleansweep software enabling a long list of experiments. I need to acknowledge Lars for his positive mood, for many fruitful discussions and the big amount of ingrate technical work we did together. Last but not least tanks Gabe, not only for some of the best (and scariest) boarding days of my life but also for all the time invested in debugging the many many baby sicknesses of our Vericold dilution refrigerator.

List of Publications

1. S. Filipp, M. Göppl, J. M. Fink, M. Baur, R. Bianchetti, L. Steffen, and A. Wallraff, *Multi-Mode Mediated Qubit-Qubit Coupling and Dark-State Symmetries in Circuit Quantum Electrodynamics*, Physical Review A **83**, 063827 (2011). DOI:10.1103/PhysRevA.83.063827 or arXiv:1011.3732

2. D. Bozyigit, C. Lang, L. Steffen, J. M. Fink, C. Eichler, M. Baur, R. Bianchetti, P. J. Leek, S. Filipp, M. P. da Silva, A. Blais, and A. Wallraff, *Antibunching of Microwave Frequency Photons observed in Correlation Measurements using Linear Detectors*, Nature Physics **7**, 154-158 (2011). DOI:10.1038/nphys1845

3. D. Bozyigit, C. Lang, L. Steffen, J. M. Fink, C. Eichler, M. Baur, R. Bianchetti, P. J. Leek, S. Filipp, M. P. da Silva, A. Blais, and A. Wallraff, *Correlation Measurements of Individual Microwave Photons Emitted from a Symmetric Cavity*, Journal of Physics: Conference Series **264**, 012024 (2011). DOI:10.1088/1742-6596/264/1/012024

4. R. Bianchetti, S. Filipp, M. Baur, J. M. Fink, C. Lang, L. Steffen, M. Boissonneault, A. Blais, A. Wallraff, *Control and Tomography of a Three Level Superconducting Artificial Atom*, Physical Review Letters **105**, 223601 (2010). DOI:10.1103/PhysRevLett.105.223601 or arXiv:1004.5504

5. J. M. Fink, L. Steffen, P. Studer, L. S. Bishop, M. Baur, R. Bianchetti, D. Bozyigit, C. Lang, S. Filipp, P. J. Leek and A. Wallraff, *Quantum-to-Classical Transition in Cavity Quantum Electrodynamics (QED)*, Physical Review Letters **105**, 163601 (2010). DOI:10.1103/PhysRevLett.105.163601 or arXiv:1003.1161

6. D. Bozyigit, C. Lang, L. Steffen, J. M. Fink, M. Baur, R. Bianchetti, P. J. Leek, S. Filipp, M. P. da Silva, A. Blais, and A. Wallraff, *Measurements of the Correlation Function of a Microwave Frequency Single Photon Source*, (2010) arXiv:1002.3738

7. P. J. Leek, M. Baur, J. M. Fink, R. Bianchetti, L. Steffen, S. Filipp, and A. Wallraff, *Cavity Quantum Electrodynamics with Separate Photon Storage and Qubit Readout Modes*, Physica Scripta **104**, 100504 (2010). DOI:10.1103/PhysRevLett.104.100504 or arXiv:0911.4951

8. J. M. Fink, M. Baur, R. Bianchetti, S. Filipp, M. Göppl, P. J. Leek, L. Steffen, A. Blais and A. Wallraff, *Thermal excitation of multi-photon dressed states in circuit quantum electrodynamics*, Physica Scripta **T137**, 014013 (2009). DOI:10.1088/0031-8949/2009/T137/014013 or arXiv:0911.3797

9. R. Bianchetti, S. Filipp, M. Baur, J. M. Fink, M. Göppl, P. J. Leek, L. Steffen, A. Blais, and A. Wallraff, *Dynamics of dispersive single-qubit readout in circuit quantum electrodynamics*, Physical Review A **80**, 043840 (2009). DOI:10.1103/PhysRevA.80.043840 or arXiv:0907.2549

10. J. M. Fink, R. Bianchetti, M. Baur, M. Göppl, L. Steffen, S. Filipp, P. J. Leek, A. Blais and A. Wallraff, *Dressed Collective Qubit States and the Tavis-Cummings Model in Circuit QED*, Physical Review Letters **103**, 083601 (2009). DOI:10.1103/PhysRevLett.103.083601 or arXiv:0812.2651

11. M. Baur, S. Filipp, R. Bianchetti, J. M. Fink, M. Göppl, L. Steffen, P. J. Leek, A. Blais and A. Wallraff, *Measurement of Autler-Townes and Mollow Transitions in a Strongly Driven Superconducting Qubit*, Physical Review Letters **102**, 243602 (2009). DOI:10.1103/PhysRevLett.102.243602 or arXiv:0812.4384

12. P. J. Leek, S. Filipp, P. Maurer, M. Baur, R. Bianchetti, J. M. Fink, M. Göppl, L. Steffen and A. Wallraff, *Using sideband transitions for two-qubit operations in superconducting circuits*, Physical Review B **79**, 180511 (2009). DOI:10.1103/PhysRevB.79.180511 or arXiv:0812.2678

13. S. Filipp, P. Maurer, P. J. Leek, M. Baur, R. Bianchetti, J. M. Fink, M. Göppl, L. Steffen, J. M. Gambetta, A. Blais and A. Wallraff, *Two-Qubit State Tomography Using*

a Joint Dispersive Readout, Physical Review Letters **102**, 200402 (2009). DOI:10.1103/PhysRevLett.102.200402 or arXiv:0812.2485

14. A. Fragner, M. Göppl, J. M. Fink, M. Baur, R. Bianchetti, P. J. Leek, A. Blais, A. Wallraff, *Resolving Vacuum Fluctuations in an Electrical Circuit by Measuring the Lamb Shift*, Science **322**, 1357 (2008). DOI:10.1126/science.1164482

15. M. Göppl, A. Fragner, M. Baur, R. Bianchetti, S. Filipp, J. M. Fink, P. J. Leek, G. Puebla, L. Steffen, and A. Wallraff, *Coplanar waveguide resonators for circuit quantum electrodynamics*, Journal of Applied Physics **104**, 113904 (2008). DOI:10.1063/1.3010859 or arXiv:0807.4094

16. J. M. Fink, M. Göppl, M. Baur, R. Bianchetti, P. J. Leek, A. Blais and A. Wallraff, *Climbing the Jaynes–Cummings ladder and observing its nonlinearity in a cavity QED system*, Nature **454**, 315 (2008). DOI:10.1038/nature07112 or arXiv:0902.1827

17. P. J. Leek, J. M. Fink, A. Blais, R. Bianchetti, M. Göppl, J. M. Gambetta, D. I. Schuster, L. Frunzio, R. J. Schoelkopf, and A. Wallraff, *Observation of Berry's Phase in a Solid State Qubit*, Science **318**, 1889 (2007). DOI:10.1126/science.1149858 or arXiv:0711.0218

18. R. Leturcq, R. Bianchetti, G. Götz, T. Ihn, K. Ensslin, D. C. Driscoll and A. C. Gossard, *Magnetic field symmetry and phase rigidity of the nonlinear conductance in a ring*, AIP Conference Proceedings **893**, 709-710 (2007). DOI:10.1063/1.2730087

19. R. Leturcq, R. Bianchetti, G. Götz, T. Ihn, K. Ensslin, D. C. Driscoll and A. C. Gossard, *Coherent nonlinear transport in quantum rings*, Physica E **35**, 327 (2006). DOI:10.1016/j.physe.2006.08.023

Die VDM Verlagsservicegesellschaft sucht für wissenschaftliche Verlage abgeschlossene und herausragende

Dissertationen, Habilitationen, Diplomarbeiten, Master Theses, Magisterarbeiten usw.

für die kostenlose Publikation als Fachbuch.

Sie verfügen über eine Arbeit, die hohen inhaltlichen und formalen Ansprüchen genügt, und haben Interesse an einer honorarvergüteten Publikation?

Dann senden Sie bitte erste Informationen über sich und Ihre Arbeit per Email an *info@vdm-vsg.de*.

Sie erhalten kurzfristig unser Feedback!

VDM Verlagsservicegesellschaft mbH
Dudweiler Landstr. 99   Telefon +49 681 3720 174
D - 66123 Saarbrücken   Fax   +49 681 3720 1749

www.vdm-vsg.de

Die VDM Verlagsservicegesellschaft mbH vertritt

Printed by Books on Demand GmbH, Norderstedt / Germany